Making Space

Making Space

Property Development and Urban Planning

Edited by Andrew MacLaran

Department of Geography, Trinity College, Dublin, Ireland

Arnold
A member of the Hodder Headline Group
LONDON

Distributed in the United States of America by
Oxford University Press Inc., New York

First published in Great Britain in 2003 by
Arnold, a member of the Hodder Headline Group,
338 Euston Road, London NW1 3BH

http://www.arnoldpublishers.com

Distributed in the United States of America by
Oxford University Press Inc.
198 Madison Avenue, New York, NY 10016

British Library Cataloguing in Publication Data
A catalogue record for this book is available from the British Library

Library of Congress Cataloging-in-Publication Data
A catalog record for this book is available from the Library of Congress

ISBN 0 340 808276 (pb)
ISBN 0 340 808268 (hb)

1 2 3 4 5 6 7 8 9 10

Typeset in 10/14pt Gill Light by Phoenix Photosetting, Chatham, Kent
Printed and bound in Malta

What do you think about this book? Or any other Arnold title?
Please send your comments to feedback.arnold@hodder.co.uk

Contents

Acknowledgements

I am grateful to the following for permission to include copyright material:

Professor Murray Wilson for Figure 2.2; The City Assessor's office, Minneapolis for the data for Figures 4.7 and 4.8; 'Axiss Australia' for permission to make use of data on the ownership profile of prime office space data in Figure 5.5; the Property Council of Australia for permission to make use of data for Figure 5.6; Irish Geography for permission to use Figure 6.5; Peter Barrow, European Photo Services, for Figures 6.7 and 6.8; The International Journal of Urban and Regional Research and Blackwell Publishing for permission to draw in Chapter 7 on an article by Moricz and Murphy (1997).

Whilst every effort has been made to trace copyright owners, this has not been possible in some instances and apologies are extended to those whose rights have unintentionally been infringed.

Andrew MacLaran

List of contributors

Carrie Breitbach is a doctoral student at Syracuse University, New York, researching economic restructuring and social change in the beef industry in South Dakota.

John Bryson is a Reader in the Department of Geography, University of Birmingham, UK.

David Laverny-Rafter is a Professor of Urban Studies at Minnesota State University, Mankato.

Andrew MacLaran is a Senior Lecturer in the Department of Geography, Trinity College Dublin, and joint director of the Centre for Urban and Regional Studies, TCD.

Pauline McGuirk is a Senior Lecturer in the School of Environmental and Life Sciences at the University of Newcastle, New South Wales and deputy director of the Centre for Urban and Regional Studies, Newcastle University, New South Wales, Australia.

Don Mitchell is a Professor in the Department of Geography, Maxwell School of Citizenship and Public Affairs at Syracuse University, New York.

Laurence Murphy is a Senior Lecturer in the Department of Geography at the University of Auckland, New Zealand.

Brendan Williams is a Lecturer in the Faculty of the Built Environment at the Dublin Institute of Technology, Bolton Street.

Preface

The idea for this book took shape several years ago, its delayed appearance being occasioned by administrative duties and a necessary subsequent quest to retrieve a degree of sanity. The text has been a collaborative effort and I wish to thank the contributors for their support, feedback and patience as it took shape. Its shortcomings, however, are entirely my responsibility.

For taking charge of the production of most of the illustrations, I am greatly indebted to Sheila McMorrow of the Department of Geography, Trinity College Dublin. Thanks are also extended to Colin MacLaran, Michael Punch and Declan Redmond for their observations on some initial drafts and to Patrick Malone for his constructive comments and our many discussions.

Andrew MacLaran
Trinity College, Dublin
November 2002

1
Creating urban space

Andrew MacLaran

This book is about building cities. It focuses on the private-sector forces responsible for their development and the arrangements put in place to guide, manipulate and control them. Perhaps, in an era of increasing 'globalization', characterized by the instant transfer of finance and information across the globe, it is no longer necessary and perhaps even unwise to differentiate between urban and rural realms, at least in the developed world. Here, capitalist market-based systems, relations of production and urban-based media have penetrated the furthest recesses of the countryside, reshaping traditional ways of life, economies, politics and culture, to the extent that it becomes virtually impossible to delimit the urban arena. Nevertheless, the designation 'urban' does still convey meaning.

Cities are fascinating because, in our increasingly urban world, they are intimately tied into almost the complete totality of human life. This is what makes them such vibrant and exciting places in which to live and also such interesting places to study. Cities mirror the character of the society that creates and sustains them. They are therefore as multifaceted as the range of social complexity which engenders them. They express the ways in which people spend their lives, make their living and use their leisure time. They reflect the specialized forms of human economic activity which take place in urban areas, the relationships and interactions which underpin them and the legal forms which support and enforce them. They express the manner in which people are organized politically and shaped ideologically. They concern cultural, religious and artistic expression, reflecting varying forms of family life and gender roles, as well as the ways in which children are raised, educated and socialized into the adult world.

Our personal images of cities may well emphasize their social aspects, simultaneously encapsulating both positive and negative elements: the alienation of meaningless working lives, yet the almost tangible energy and stimulation of life at the economic and cultural heart of things: the congregation of people and the superficial depth of personal interactions, yet the 'buzz' or 'craic' of living to the pulse of modernism; the anonymity of the crowd, yet the behavioural freedoms this brings, unfettered by informal restraint of cleric, neighbour or even family ties; the promise of economic opportunity and social fulfilment, tarnished by realities of housing affordability crises, traffic congestion, fear of crime and for personal safety.

Obviously, then, cities are more than just buildings. Yet, a powerful and highly durable component of our image of urban areas is constituted by the built environment itself. Cities such as New York, San Francisco, Sydney, London and Paris are instantly identifiable by their unique skylines created by the massing of buildings, especially in their central business areas, or by the presence of particular landmark buildings. Even on a more intimate scale, it is hard to think of our home neighbourhoods without conjuring up an image of streets and buildings as well as the ties of family and friendships that make those places significant. Buildings themselves become endowed with personal meaning.

Cities comprise a plethora of functions, broadly relating to the productive, distributive, reproductive and exchange operations of our social system. These each have different requirements and preferences with regard to building type and urban situation. Locational competition between those functions, with their different degrees of economic power, together with the operation of planning systems which have often favoured mono-functional zoning, create an urban landscape in which different functions become inscribed in geographical space. These specialized functional areas, industrial, commercial and residential, are linked to one another through transport and communication infrastructure, facilitating the movement of goods, information and people, permitting the daily assembly of the labour force at its place of employment and its return home to recuperate after work. It is 'built space' which provides for the accommodation of all these activities and, if anything truly typifies the 'urban' and dominates our image of cities, it is probably their built environment. The chapters that follow are concerned with the making of that urban space, the product of a distinct process.

The urban fabric embodies considerable investment of materials and labour power in its construction and this can usually be written off only slowly, typically over many decades. Change is therefore normally also slow. Occasionally, whole new towns may be created but, generally, the vast majority of buildings are inherited from previous decades, sometimes from previous centuries. Nevertheless, cities do change. Buildings are adapted to accommodate different functions and land is upgraded to provide for more profitable uses. Areas whose functions have been overtaken by economic or technological change become redeveloped, while peripheral expansion accommodates growth.

Figure 1.1 Les Arènes de Picasso, Marne-la-Vallée, France. As a non-market-related sector, social housing may exploit the freedom to be architecturally adventurous

The public sector often plays a direct role in the development of urban space, particularly of those elements which may be too risky for the private sector to undertake or which lie beyond the logic of individual capitalists to provide, such as transport infrastructure or housing for the poorer classes. However, it is normally the private sector that is responsible for the greater part of real-estate development. Therefore, it is the role of the private-sector property development 'industry' which forms the predominant focus of this book. The operations of this industry help to shape the built environment, influenced by powerful property interests, commercial companies and investment institutions. Indeed, as Feagin (1983, 3) has emphasized, urban space is not simply a neutral container of societal functions:

> Cities under capitalism are structured and built to maximize the profits of real estate capitalists and industrial corporations, not necessarily to provide decent and livable environments for all urban residents.

However, the industry's activities are not unfettered. The state has almost invariably sought to intervene in the process of creating the built environment through urban planning systems which seek to control, shape and influence the outcomes.

Chapters 2 and 3 provide the context for understanding the creation of the built environment. Chapter 2 describes the major agents and actors involved in the property development 'industry'. It investigates their motivations and modes of operation, examining the relationships between them and their struggle to share in the profits which can accrue from development. It discusses the importance of location on the potential profitability of commercial property development and the impact which this can have on the urban landscape. It also reviews the existence of development cycles and their impacts on the property sector.

Chapter 3 addresses the topic of urban planning, examining its operations and briefly outlining its philosophical and ideological bases. Urban planning involves the intervention by the state, usually at local level, in the process of property development. Thus, it becomes an important influence on development by establishing the context within which the creation and redevelopment of the urban environment takes place. It may guide, support, promote or hinder development. It may regulate the way in which development takes place and seek to control its potentially adverse consequences by protecting environmental resources and conserving valued natural or built heritage. In so doing, it seeks to achieve certain desired spatial or social arrangements which otherwise may not be attained. Traditionally, intervention has involved regulating the uses to which urban land and buildings are put and controlling the character of development by determining maximum development densities (plot ratios) and building height. It may also exercise control over construction quality and vet the architectural design of schemes.

However, as urban planning comprises an activity of the state, in order to understand why the state intervenes at all in the creation and redevelopment of urban space requires some consideration of the role of the state in capitalism. Because there are several conceptions of state functions, there are necessarily several distinct interpretations of the role of urban planning and these require brief review.

The chapter draws particular attention to the manner in which strategic urban planning goals and the practice of urban planning have tended to change in recent years. As the operations of the state itself have tended internationally to become more 'entrepreneurial' and facilitative of the interests of capital, so changes in urban governance have frequently led to the transformation of urban planning towards more directly pro-active modes of operation supportive of private-sector development interests. Consequently, urban planning has encompassed property-led urban regeneration policies promoted and often generously supported financially by the state, as it has attempted to harness the forces of property development in strategies of urban growth and renewal, thereby enhancing and transforming the city's image and boosting urban competitiveness for inward investment. Thus, the conception of urban planning as an obstructive hurdle to development has become largely displaced in the western world over the past 20 years as it adopted increasingly entrepreneurial approaches that provided selective inducements to tempt developers towards certain courses of action rather than others. Increasingly, the carrot has tended to replace the stick.

The broad introductory discussions of Chapters 2 and 3 create a framework for understanding the nature of property development and the functions of urban planning. However, unique national and geographical situations arise which influence the precise outcomes. It is these twin elements of planning and development which are taken up in the six case-study chapters that follow (Chapters 4–9), adding flesh to the skeleton of generalization provided by the earlier chapters. Six cities have been selected for inclusion. Two are North American, two Australasian and two North European. They are Minneapolis (Minnesota), Sioux Fall (South Dakota), Sydney (New South Wales), Auckland (Aotearoa/New Zealand), Birmingham (England) and Dublin (Ireland). They share a certain heritage with respect to planning practice and a similar ethos concerning private property rights.

The selected cities are not of first rank on the global stage. There are two main reasons for this choice. First, world cities such as London and New York are unusual in many respects, being more closely tied into global events and processes compared to those of secondary or tertiary status which have more intimate regional ties. Their stock of buildings tends to reflect this. For example, London's huge stock of office property comprises around 60 per cent of the UK total. The interest of property developers and investors also tends to be drawn from a global scale. Second, both New York and London have already been the focus of some excellent and readily available reviews, notably that of Fainstein (1994).

The aim of these case-study chapters is to investigate the ways in which urban planning in different national contexts has sought to influence the results of the property development process and the way in which the modes of operation of planning have developed in order better to effect such influence. This strategy argued strongly in favour of cities which were more typical of their societies, where the roles of the regional economy and local planning systems were more evident, yet which were still of a sufficient scale to experience significant property development pressures.

The first of the case studies is based on Minneapolis, Minnesota. Its focus is somewhat different from the remaining five cities as the chapter sets out to demonstrate, in the context of a market-dominated economy, that urban planning historically was a creature of private-sector needs. It aims to contradict the notion that planning represents some radical challenge to private property rights or a conspiracy to thwart the interests of the development sector. It shows instead that urban planning was in fact conceived in response to the requirements of downtown property and business interests and that it engaged from its earliest days in a highly entrepreneurial role which facilitated those interests. This conferred immense benefits on those with property, while imposing enormous real costs on those without.

The second of the case-study chapters focuses on Sydney, New South Wales, as it endeavoured to establish itself as a business centre of global significance. In examining the transformation of its central business area, the chapter addresses the problems which result from the fragmentation of urban planning authority between regional and local levels of control. With a predominantly state-level planning authority which was highly supportive of the property development sector, it reveals the difficulties that locally-based planning experienced in balancing the demands of developers with the conflicting views and objectives of community groups about the manner in which the central area was to develop.

Dublin, Ireland, is the subject of the third case study. It reviews how traditional urban planning, based on land-use zoning and development control, proved largely ineffective in dealing with the intensity of inner-city decline and dereliction prevailing during the 1980s. It examines the national government's response to such shortcomings, notably the creation of financial incentives for designated urban renewal areas and the establishment of special-purpose agencies to promote private-sector property-based renewal. This engendered a marginalization of traditional urban planning and led to a rethinking of the operations of urban planning in the city.

The changing dynamics between the various agents involved in urban development are investigated in the study of Auckland, New Zealand. It examines the transformation of the city's central business area during the course of a major office development boom, the subsequent boom in the construction of high-rise apartments and, finally, the impact of partnerships between the public and private sectors in the creation of 'spaces of consumption and spectacle'.

The chapter on Birmingham, England's second city, examines the way in which national planning policy has been interpreted locally. It reviews the manner in which the city has sought, by taking cognisance of the needs of property developers and investors, to employ the transformational power of the private-sector property development industry to recreate itself as a city of the twenty-first century.

The final case study reviews the transformation of Sioux Falls, South Dakota. It takes up the theme of urban 'boosterism' and examines the way in which the forces of property development have been harnessed locally in a strategic project to transform the city from a provincial north-plains meat-packer to a financial business centre of national significance. It shows how planning and property development have been used by the business élite in a scheme to re-create the urban landscape as one which embodies the culture of economic expansion, in which the urban landscape itself has become transformed into an icon of growth. However, as the authors are at pains to show, in this process of reshaping the city, clear winners and losers have inevitably emerged.

It should become apparent that, despite their geographical distance from one another, there is considerable similarity between the case studies in terms of the problems experienced, the planning processes adopted and the resultant outcomes.

2
Masters of space: the property development sector

Andrew MacLaran

This chapter reviews the operations of the property development sector, the industry largely responsible for making and re-shaping the built form of our cities. It focuses on private-sector operations, investigating the roles of the major actors and agents involved in the process of development, their motivations and the relationships between them. It examines the industry's propensity to undergo cycles of boom and slump in the scale of development activity and reviews some of the consequences for the development sector itself.

2.1 Development operations

There are probably few individuals or institutions that have attracted such an intense degree of opprobrium as have property developers, suffering the vitriolic admonishments of journalists, community groups and conservationists alike. However, like other industries, property development is simply engaged in the production of commodities. In this case, the commodities comprise the various types of building (e.g. housing, offices, shops, hotels, factory units and warehousing) developed to meet a demand from users. Built space is therefore a prerequisite for the accommodation of vital social functions, as Boddy (1981, 281) has observed:

> Property development and investment equips space and creates physical infrastructure so as to facilitate the circulation of capital in its various forms, the exchange of information, and the physical and legal transactions on which commercial, administrative, governmental and financial functions are predicated.

Although the built environment is relatively fixed, each building physically embodying a considerable investment of capital and labour power, it does slowly change. As new types of function appear and as others die out, or as functions alter their locational preferences or building requirements, these different demands become reflected in changes in the built environment. The city expands laterally and adapts internally to the changing demands of those requiring built space. Thus, urban change is typically driven by the dynamics of businesses and households, the outcomes being mediated through land and property markets. This is well expressed by D'Arcy and Keogh (1997, 690):

At its simplest, urban economic activity generates requirements for land and property which might be met by the existing stock of buildings or through new development. These requirements are mediated through a property market process which, amongst other things, determines property values, allocates space in buildings between competing uses, and stimulates the production of new space through development and redevelopment.

Property markets comprise 'a network of rules, conventions and relationships which collectively represent the system though which property is used and traded' (Keogh and D'Arcy, 1999, 2408). Thus, the legal basis governing the way in which property is held and used, the mix of market and non-market mechanisms and the degree of professionalization of the property sector are features differentiating property markets internationally and through time.

2.1.1 Peripheral expansion

At the urban periphery, the property development sector frequently operates on a large scale. It acts as the 'cutting edge' of the advancing city, annually converting tens and possibly hundreds of hectares of agricultural land from land-extensive low-value rural functions to land-intensive higher-value urban activities (see Figure 2.1). Here, the largest users of land have historically been residential and industrial functions, with shopping malls and office parks making their mark more recently.

Figure 2.1 *Sydney and the Bush. Residential development at the cutting edge of the city, Campbelltown, Sydney*

2.1.2 Internal modification

Within the city, the scale of the industry's operations is generally less extensive. Here, the industry is concerned with the reconditioning of buildings, their conversion to new uses and the redevelopment of land. Although each site might be quite small individually, the aggregate effects of numerous small-scale central-city redevelopments can be visually dramatic. Figures 2.2 and 2.3 reveal the transformation of Sydney's

Figure 2.2 *Central Sydney, 1955, an essentially medium-rise central business district*

Figure 2.3 *Central Sydney, 2000, the redevelopment of numerous sites resulting in the transformation of Sydney's skyline*

Figure 2.4 *The disused Royal Victoria Dock, London*

central business district (CBD) and Circular Quay waterfront between 1955 and 2000 as a plethora of small sites underwent site assembly and piecemeal redevelopment.

In many cities, larger-scale coordinated redevelopment projects of former industrial sites and disused docklands have contributed to the overall impact. These have often been promoted and financially supported by the public sector (see Brownill, 1990; Healey *et al.*, 1992; Healey, 1994; Imrie and Thomas, 1999). Such interventions might involve central or local governments providing special grants or tax inducements to developers, or the liberalization of planning regimes to stimulate development by cutting 'red tape' in the planning process. It might also involve creating special-purpose redevelopment and renewal authorities, such as urban development corporations. These might be endowed with urban planning powers to streamline planning controls and possess the operational and financial powers to undertake infrastructural developments, acquire land, reclaim it and dispose of it at a subsidized cost to private-sector developers.

2.2 Modelling development: a property 'industry'?

In order to understand how property development operates and the ways it affects our cities, it is important to examine the character of the industry and its driving force. It should then become evident that the development of buildings does not occur as a simple reflex reaction to changing user demand. It is also governed by the profit-seeking criteria of the numerous elements of capital involved in the creation of the built environment. Development will take place only if the criteria for engagement for each of these interests are met. Their separate 'terms of engagement' are established through negotiations in different markets: between developers and landowners to establish the price (and other conditions) of access to land for development; between developers and their financiers to determine the cost and conditions of financing; and between developers and building contractors to set construction cost. If the criteria for engagement of any interest are not satisfactorily met, development is unlikely to happen.

Thus, the property development sector does not exist as an identifiably separate and coherent industry. Rather, the 'industry' comprises an assemblage of different types of capital, interests and functions. In an excellent review of different models of the development process, Healey (1991) has noted four approaches which provide rather different perspectives:

1. *Equilibrium models*, derived from neo-classical economics, which assume that development activity responds simply to signals measuring effective demand, reflected in rental levels and yields (see below).
2. *Event-sequence models*, which focus on the sequencing and management of stages in the development process.
3. *Agency models*, which concentrate on the actors involved and the relationships between them.
4. *Structure models*, that focus on the forces which organize the relationships of the development process, drive its dynamics and are strongly grounded in urban political economy.

There are merits to each of these, but the procedure adopted in this chapter draws on Malone's (1985) model which focuses on the archetypal roles and relationships which characterize private-sector property development (see MacLaran, 1993). Certain archetypal interests are associated with each function. These are set out in Figure 2.5. Briefly, these include landowners, developers and investors in property. These may be referred to as 'property interests' because their association with development and the built environment depends on their possessing certain 'property rights'. There are also construction and financial interests, representing respectively industrial and finance capital. All are in competition with one another for shares in the profits that can accrue from property development.

Admittedly, this is a highly over-simplifying approach which loses the complexity and detail provided by models of the development process or of the institutional relationships comprising the development system (see Barrett et al., 1978; Ambrose, 1986; Healey, 1992). It also lacks the depth of Ball's (1983, 1998) 'structures of provision' approach which examines the 'contemporary network of relationships associated with the provision of particular types of building at specific points in time' (Ball, 1998, 1513).

However, the approach adopted here has the merit of both simplicity and clarity and, because it directs attention towards archetypal roles rather than at specific actors, institutional arrangements, legal and operational contexts, it possesses wide geographical and temporal applicability. It can encompass varying institutional arrangements and changing organizational structures, embrace the changing mix of actors within each of

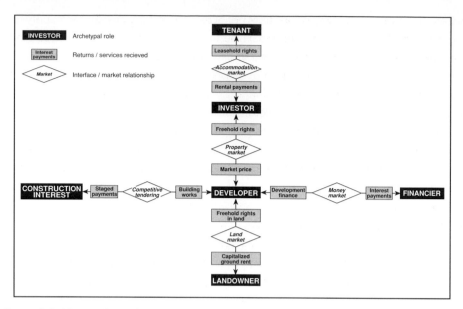

Figure 2.5 *Major archetypal private-sector roles and relationships in the property sector*

the roles, their differing strategies and criteria for engagement in the development process as well as the changing relationships between them. It therefore provides a useful skeletal foundation upon which more detailed and specific local understandings can be constructed.

In brief, the model shows that the development role is key to the whole operation, initiating the process on the basis of a predicted demand for the product: a particular type of building at a specific location. Perceiving an opportunity to profit from this, the developer is responsible for putting the scheme together. It is the developer who negotiates with landowners to acquire development rights to sites. These may be purchased freehold or as a leasehold interest for a specified period (perhaps 125 years). Developers arrange short-term financing for site acquisition and to fund construction. They also commission architects, quantity surveyors, chartered surveyors and planning consultants to devise a scheme within certain cost constraints that will be acceptable to planning authorities, yet will produce a profit. It is also the developer who engages the builders and appoints estate agents to seek suitable tenants and/or purchasers for the completed building. Thus, developers should perhaps be regarded as the impresarios of the built environment. Driven by the search for profit, if things should go wrong, it is the developer who also carries the risk, albeit often risking funds that have been borrowed. The collapse of major commercial developers such as Olympia and York (O&Y) and Rosehaugh testifies to the fragility of prolonged success.

The economic returns to all these interests are ultimately derived from the occupiers of finished buildings. These may be either in the form of a flow of rental payments to the long-term investor-landlord, or as a single lump sum payment represented by the building's market price (capitalized rent). Thus, the returns to landowners, developers, financiers and construction interests represent advance shares of this value.

The model depicts only archetypal roles. In reality, a single individual, company or institution can occupy several roles simultaneously and perform numerous functions in the development process. For example, if a construction company uses its own funds to redevelop its own site for offices, carrying out the building work itself, it would be acting in the roles of landowner, developer and construction interest. If it were to retain the development, either for its own managerial accommodation or rent it out to tenants, it would further adopt the roles of occupier or investor. Similarly, if an institution such as a life assurance company were to undertake an office development using its own development finance and subsequently transfer the property into its investment portfolio, it would take on the roles of developer, financier and investor.

Furthermore, the nature of the relationships between the archetypal interests also varies considerably. The simple market links depicted in the model disguise a diversity of possibilities. For example, between developers and landowners, the simple market relationship of outright purchase of development rights ignores the possibility of the landowners entering into a joint development partnership with the developers to share in development profits. Similarly, with respect to the raising of development finance, the securing of capital in return for interest payments is but one of many methods by which finance can be secured (see Ratcliffe and Stubbs, 1996). For example, in return for a reduced rate of interest, financiers might seek a degree of equity participation in the project, or the developer instead may raise funds through a stock-market share issue.

Additionally, it should be remembered that numerous professional advisers and agents (e.g. architects, structural engineers, planning consultants, project managers, solicitors and estate agents) are also involved in the development process. As they generally operate on a percentage-related fee-for-service basis, rather than their competing directly for development profits, they are excluded from the model. However, their absence does not signify their unimportance in shaping the manner in which development takes place, its character or location.

2.3 Development roles and relationships

To understand the property sector, it is important to comprehend the operations and motivations of the primary interests. This requires us to add some flesh of reality to the bones of the model. Probably the most appropriate place to start is with the development function, which can be seen from the model to occupy the pivotal role.

2.3.1 Developers (commercial capital)

Property developers exercise the right to develop land, transforming it from one state to another by arranging for preparatory site-works and the construction of a building. Private-sector developers comprise a highly heterogeneous group. They range from individual entrepreneurs, partnerships and private businesses, to large international companies which may be quoted on the stock exchange. They include property investment trusts and the property development departments of transnational financial institutions, notably the life assurance companies. In order to secure a flow of building work, construction companies have also taken on the development role, frequently by establishing separate development subsidiaries.

It is common to distinguish between types of developer (see CALUS, 1979). The prime concern of the *trader-developer* or *developer-seller* is to organize and manage the development process then, upon completion, sell the scheme to a long-term investor for a development profit. In contrast, *developer-investors* undertake the same development operations with the intention of retaining the completed scheme and receiving the flow of rental income from tenants. Similarly, *institutional investor-developers* normally undertake development in order to retain the property as a long-term investment. However, from the perspective of development motivation, categories are not watertight. While smaller-scale developers are more likely to be traders and larger property companies often act out of an investment motive, in reality, motivation can switch from one period to another between generating immediate profit or that of long-term investment. Furthermore, even the long-term investors have tended to trade or 'turn over' a proportion of their property investments annually in order to keep their portfolios balanced and up-to-date (Debenham, Tewson and Chinnocks, 1989; Key, Espinet and Wright, 1990).

While some developers will restrict their field of operations to a single location, others become active in a number of cities, regions or even different continents as development opportunities arise. Thus, during the UK development boom of the 1980s, overseas property companies from the USA and Japan were important players, often bringing with them rather different attitudes towards risk and challenging local operational traditions (Healey and Barrett, 1990). Some major multi-locational developers may operate through local partners better attuned to the vagaries of the local planning context, the conditions of availability of land, with a better understanding of factors relating to demand for the finished product (Logan, 1993). Furthermore, while some developers operate in a number of different property sectors, others specialize in one or two, be it industrial property, retail schemes, offices or housing. So, even if we exclude the huge variety of developments undertaken by owner occupiers, the range of developers acting in a single property sector can be very diverse.

2.3.2 Landowners (landed capital)

A prerequisite for any development is the availability of land. However, land is an object of property rights[1] conferring certain powers over its use. The nature and extent of these rights are legally defined and the variety of systems of tenure is considerable. These prescribe the rights to own, occupy, develop, lease, bequeath and sell the property (Kivell, 1993). For our purposes, the most important distinction is between 'freehold' ownership in perpetuity and 'leasehold' interests that are limited in duration, commonly 125 years, though historic leaseholds may extend to as long as 999 years.

On the basis of their exclusionary right, landowners, representing a fraction of property capital, are able to extract payments for the use or development of land by others. These rights in land can be traded, its rental value being capitalized in the form of a sale price (i.e. at a rate of, perhaps, 10 or 20 times the prevailing annual rental value). In Figure 2.5, the landowner is the interest which gives up certain property rights in land to a developer, generally in return for a lump sum payment, although a profit-sharing agreement may also be entered into with the developer.

The maximum price a developer can afford to pay for land is a 'residual' value. This is determined by the estimated value of the completed project minus the cost of site preparation, construction, financing costs, professional fees and the developer's profit (see Cadman et al., 1991; Adams, 1994). Of course, this is something of a gamble, as it relies on the accuracy of prediction of many factors as yet unknown. If property values fall during development, or if costs rise, it can spell ruin, as many developers have found to their cost.

Landowners range from individual persons and small businesses to companies with wide regional, national or international interests whose strategies with regard to profiting from the ownership or disposal of specific sites will be set within a broader business context (see Healey, 1994; Ball, 1998). Private owners also include insurance companies, pension funds and investment companies. Because of their diversity, landowners have a variety of goals, which may not be simply that of maximizing profit. Even among businesses, there may be numerous strategies associated with land ownership, sharply distinguishing between industrial owners, using land as one of a number of factors of production and perhaps retaining fallow land for future plant extension, and the strategies of land dealers and speculators, property developers and financial owners. A discussion of landowners' perspectives and motivations is provided by Massey and Catalano (1978) and by Adams (1994).

The significance of traditional landowners such as the Church, the Crown and the landed aristocracy can be considerable in some cities, especially in Europe. In London, for example, the urban estates of the Portman and Cadogan families, the Duke of Portland (Howard de Walden) and Westminster (Grosvenor Estate) remain largely intact, the latter also owning substantial land in Chester and Liverpool (Marriot, 1967; Green, 1986; Kivell, 1993).

National and local governments and the quasi-public/semi-state sector, such as nationalized industries, are also important landowners. The extent of such public ownership varies internationally, generally being least significant in market-dominated economies such as the USA and more common in European states because of historically differing political philosophies and agendas. There are periods in which land might be taken into public ownership ('nationalization') or when ownership rights become increasingly constrained through planning legislation or the taxation of 'betterment' ('unearned' gains arising to landowners as a result of urban expansion and public infrastructural developments). These can alternate with phases of land 'privatization' in which, at the behest of national or local states, policies encourage the disposal of public-sector assets such as land or promote the establishment of joint development ventures with the private sector to undertake development of sites in public ownership. For example, at its strategically located London railway termini, British Rail (BR) undertook major property developments in association with a number of developers, such as Greycoat, Rosehaugh-Stanhope and Speyhawk, BR making St£319 million from land sales and leases in 1989–90, (Kivell, 1993). However, in selling sites, the public sector often possesses far broader objectives than merely maximizing the sale price, so the disposal of public-sector land has often been effected through competitions between interested parties to determine the most 'suitable' developers.

A landowner's capacity to gain advance shares in the potential profits from development is determined by the balance between the supply of sites and the strength of demand. All sites are unique in terms of their absolute location. But they also possess important differentiating characteristics in terms of their relative location with respect to the city's prime retailing or office districts and with regard to transportation infrastructure. They may also be further differentiated by their zoning status in urban plans, controlling the scale of development and permissible functions.

Two additional factors are significant in determining the balance of advantage in negotiations between landowners and developers. First, there is the question of the extent of the land holding or degree of spatial monopoly that the landowner possesses. Although ownership of a specific site confers a monopoly right over the site itself, land ownership in most cities is so fragmented that a wider monopoly is unlikely. However, only a limited number of sites are on the market at any time.

Second, there is the degree to which the site owner is aware of its development potential. Those well advised about market conditions are best able to reap a high price. Thus, in the City of London during the mid 1980s 'many nonfinancial landowners, such as livery companies, were insisting on a share in development profits on a geared ground-rent basis; freeholders were demanding as much as 20–40 per cent of the rack rent on the completed buildings' (Pryke, 1994a, 247). On the other hand, 'asset stripping' operations can result from a lack of awareness of the redevelopment value of sites where companies fail to include them in their accounts at full market value. The companies may then be acquired cheaply and the assets sold off at substantial profit (Goodchild, 1978). Similarly, an industrial company which moves operations to suburbia may decide to relocate with little regard to prevailing conditions in the development cycle and irrespective of whether site values are high or low. In a period of development inactivity, site disposal may prove highly problematic.

Among those owners who do recognize the redevelopment value of their sites, it is the expectation of a certain price level which influences the type of development that is economically feasible; high land costs being most readily borne by high-grade retailing, office, hotel, leisure and luxury-residential functions. Through time, the capacity of these functions to out-compete one another will vary. For example, during quiescent office markets, potential office development sites might be put to alternative uses such as luxury residential development, as happened during the early 1990s in central Sydney, Auckland and in the areas fringing Dublin's office core.

The state itself can play a significant role in determining the supply of land for development, as several of the case-study chapters demonstrate. Indeed, public-sector activities can be of such importance that they may even influence the structure of the development sector itself (see Healey and Barrett, 1990; Healey, 1994). This can involve the disposal of sites in public ownership to private-sector developers. More indirectly, the potential development profitability of sites can be enhanced by the relaxation of planning restrictions, the availability of direct subsidies or tax abatements for developers and property investors, or by the provision of infrastructure.

Governments have also on occasions sought to recoup for the wider community some of the increase in land values arising from development. The UK's post-World War 2 application of a development charge, a capital gains tax, a betterment levy, the development gains tax and the development land tax all met with limited success (see McKay and Cox, 1979; Goodchild and Munton, 1985). The Labour government's development charge (1947) was so penal that it created a serious shortage of development land, while the betterment levy (1967) was so complex that its abolition in 1970 by a Conservative government was a foregone conclusion.

In France, vigorous state intervention in land markets to facilitate comprehensive development planning at the urban periphery date from 1958 with the creation of urbanization priority zones (ZUPs – zones à urbaniser à priorité) (see Kivell, 1993). These facilitated land assembly over large peri-urban tracts, the land being purchased at the average market price prevailing during the previous five years, normally that for agricultural land. Similarly, the deferred development zone (ZAD – zone d'aménagement déférrée) was created in 1962, typically to control land speculation near new towns where very long-term planning was required. This gave the public sector the pre-emptive right to purchase land for 8 years, later 14, at a price equal to that prevailing during the year prior to the original ZAD designation. By 1977, over 500,000 ha. had been designated in ZADs, acting as a potential discouragement to land speculation.

2.3.2.1 Site assembly and site assemblers

Because access to land is so crucial to any development, it is worth considering this at greater length. At the urban periphery, considerable speculation and dealing in land often takes place prior to construction (Ball, 1983). This may involve the purchase of farmland not zoned for urban development ('white land'), especially in the vicinity of planned urban by-passes or motorway interchanges. A subsequent re-zoning permitting urban development can create spectacular profits from the consequent rise in land values. Many developers therefore engage in trading land, buying sizeable tracts, splitting them into development parcels and selling them on to others at a considerable mark-up. As land is such a key raw material, especially for housing developers who engaged in geographically extensive low-density urban development, and considerable time is normally required to bring a development from land-acquisition stage to completion, 'land banking' is commonly undertaken. Residential developers typically hold a three- to four-year supply of development land, perhaps with additional white land held for longer-term speculative gain (Smyth, 1982). These reserves can be owned outright or, especially in the case of white land, retained by conditional contracts involving 'options to purchase' (Goodchild, 1978; Ball, 1983).

Physical characteristics are reflected in the calculation of site value as small and irregularly shaped plots are more costly to develop. The economics of construction also favours large sites for property development. Unit costs may be reduced and larger schemes can provide the opportunity to exert greater control over externalities (Lamarche, 1976). However, as large new schemes become inserted into historic centres where land ownership and development was traditionally based on small individual plots, the sheer scale of much modern development alters the grain of the townscape and is often unsympathetically intrusive (see Figure 2.6).

The commonly fragmented nature of land ownership in city centres means that large-scale redevelopment may require the assembly of a number of separate plots. During site assembly, developers will frequently operate surreptitiously, purchasing properties quietly and at arm's length, perhaps using intermediaries. This helps to obscure the redevelopment potential of the targeted sites so that the owners of neighbouring properties remain unaware of what is happening. In this way, adjacent plots may be acquired cheaply at current use values rather than at prices reflecting their far higher redevelopment value.

Figure 2.6 *The Setanta Centre, Dublin. The scale of modern developments can be intrusive on historic townscapes*

Not only is land ownership in the city commonly highly fragmented, but each parcel may incorporate numerous layers of ownership, where a whole series of sub-leases may have been granted historically. The legal complexity of gaining clear ownership title to enable redevelopment can therefore take years (Goodchild, 1978). For example, in central Dublin, much of which was originally developed in the eighteenth century on the basis of leasehold titles of 999 years, property rights are often legally highly complex. It took the British property company MEPC some ten years to assemble a 1.6 ha. (4 acre) site in the office core on St. Stephen's Green, plus a number of years subsequently to sort out all the legal problems to provide clear title before it could apply for planning permission for redevelopment (MacLaran, 1993).

The problematic nature of site assembly for redevelopment provides an opportunity for yet another interest to insert itself between the landowner and the developer. Site assemblers, who possibly lack sufficient financial power or backing to undertake the development themselves, seek a short-term gain by identifying sites that have the potential for redevelopment within the next few years. They set about putting together at least the nucleus of a development site from smaller packages, profiting by buying properties at existing use values rather than at prices reflecting the full redevelopment value. Like the developer, the site assembler may purchase properties outright. Alternatively, an 'option to purchase' the property may be arranged. This involves an agreement to buy the site at a future date for an agreed amount, normally somewhat higher than the existing use value and a non-returnable deposit of anything between 10 to 25 per cent of current value might be paid to the owner. It gives the site assembler time to put a package together by completing the acquisition of remaining plots (by purchases or gaining site options) and to enter negotiations for the sale of the assembled sites to potential developers.

Site assembly difficulties have often encouraged developers to undertake office developments at the margins of the central business district (CBD), especially at times of development boom when riskier locations appear more feasible. Here, land in industrial or institutional uses can deliver a large redevelopment site by a single transaction (see Figure 2.7). A large redevelopment scheme can have a significant environmental impact locally, particularly in a 'secondary' area where it can 'lift' the whole character of its surroundings. If it is sufficiently large, it can 'stretch' the zone of prime rental values outwards to incorporate the new development into the core. However, it creates a reduction in industrial functions around the city centre and negative consequences for blue-collar inner-city communities.

A further consequence of the length of time required for site assembly is the tendency to suspend building maintenance of those properties destined for demolition (see Figure 2.8). Site assembly may itself be a lengthy process, while securing planning permission and awaiting signs of user demand may occasion further delays to redevelopment. The resultant dereliction can therefore be long term.

Figure 2.7 *Demolition of industrial buildings near the central business area provides a substantial site for office development*

Figure 2.8 *Prolonged site assembly and suspension of building maintenance can lead to dereliction*

2.3.3 Development financiers (finance capital)

Property development ties up considerable quantities of capital in a relatively fixed form for a long time and the 'industry' is normally highly 'leveraged', being heavily dependent on borrowings. The development function itself is relatively short term and the capacity of developers to undertake a development depends on their ability either to draw on their own financial resources or to raise funds from money markets. Typically, 'bridging' or 'development' finance will be arranged for up to five years. With a 'build-for-sale' development, the interest charges are often 'warehoused' or 'rolled up' until the development is completed. These are then added to the outstanding principal and, once long-term funding has been secured on the completed development (see below), the short-term financing is repaid.

Numerous sources of finance exist and have been discussed by Ratcliffe and Stubbs (1996). Banks are the most common source of short-term loans. Commercial banks tend to advance loans at 2–4 per cent above base rate but scrutinize the complete asset base of borrowers, requiring greater security than that represented by the development scheme alone. In contrast, merchant banks tend to have recourse only to the collateral embodied in the development project itself. They are also likely to be more receptive to riskier ventures and more innovative in arranging funding syndicates to finance large developments. But these advantages come at a cost. Interest rates are higher, at 4–6 per cent above base rate, with some loss of development value possibly occuring through 'equity participation' wherein the merchant bank takes an ownership share in the building.

Instead of drawing on bank finance, developers may enter into a complete funding arrangement with an institution. Here, the development is sold on to a long-term investor, such as a life assurance company or an investment bank, which also funds the project over the short term by advancing bridging finance at a preferential rate. Alternatively, a property company might raise working capital without increasing its debt by placing equity shares for sale on the stock market. Already-listed companies can raise additional capital on the stock market through a 'rights issue' of further shares to existing shareholders at a concessionary price.

The balance between equity and debt financing is crucial. As Adams (1994, 58) has noted:

> Developers who rely too heavily on equity finance may lose control of particular projects to the equity partner or even face hostile takeover bids, if too many voting shares are issued on the Stock Exchange. However, in troubled times, equity finance may be more advantageous to a developer, since equity lenders share the risks of development and may have no automatic right to repayment. In contrast, developers who rely too heavily on debt finance, in particular on bank loans, may face a liquidity crisis, if interest rates rise steeply or unexpectedly.

While individuals and smaller property companies may lack the capacity to undertake large-scale site assembly from their own resources, larger developers might fund site acquisition from their own resources, drawing on income from existing investment portfolios. This avoids problems arising from increases in interest rates, which can seriously affect the holding costs of a scheme. Nevertheless, even for larger commercial developers, construction is normally undertaken on borrowed funds. However, the property-development arms of large financial institutions, such as insurance companies, would be well capable of also funding construction.

The returns from the development process accruing to the financial interests depend on their degree of involvement. For example, financiers may simply receive their returns in the form of interest payments. The cost of borrowings are related to market interest rates, reflecting the balance between the supply of and demand for credit, and the lending institution's evaluation of the project's riskiness. Alternatively, financiers may engage in profit sharing arrangements with the developer by 'buying into' the project.

Through the latter half of the twentieth century, different types of financing emerged. In the UK during the immediate post-war period, it was common for institutions to grant mortgages to developers in which the interest payments were rolled up until the development was complete. The developer benefited from being able to operate on low-cost financing, with payments commencing only once a flow of income from tenants or a capital gain from its disposal had been secured (see Ratcliffe and Stubbs, 1996).

In the closing decades of the twentieth century, the international trend towards financial liberalization and the increasingly global operations of finance capital provided developers with a far wider range of credit sources than previously. Together with the favourable global macroeconomic environment, this created more competitive borrowing terms for developers which facilitated a global property boom (Ball, 1994). Furthermore, the financial sector readily increased its commitment to commercial real-estate lending in the 1980s because of poor results from other lending areas, such as to less-developed countries (Leitner, 1994). By 1990, bank lending against commercial real estate topped $400 billion in the USA alone as commercial banks plunged into an 'orgy of commercial construction lending' (Warf, 1994, 315; see also Fainstein, 1994). Similarly, in the UK, total bank lending to property companies (including short-term, medium-term and, by default, long-term funding as developers failed to make loan repayments) rose from around St£10 billion in 1986 to St£41 billion in 1991 (Adams, 1994).

However, because property development is just one of many types of short-term borrowing, the conditions under which the financial interests become engaged in the process of development are determined largely by what is happening outside the property sector. There is therefore considerable truth in Richard Peiser's (1990, 498) observation that 'bankers determine what gets built, since developers depend on them for financing'.

2.3.4 Building contractors (industrial capital)

It is the building contractor which constructs the building, frequently engaging a plethora of sub-contractors for the supply of specialist services and materials, as the site notice boards of development schemes illustrate (Figure 2.9). The construction interest does not constitute a 'property interest'. Rather, it comprises a form of industrial capital. Its return from the development process is as 'profit of enterprise' generated on the basis of the difference between the contract price for building works and the costs incurred for building materials and labour power employed in the scheme. The rate of profit deemed acceptable will therefore be established in relation to prevailing levels of profit in the industrial arena rather than those pertaining to the exercising or trading of property rights.

Figure 2.9 *Sub-contracting is a common feature of the construction sector*

A variety of arrangements apply to the relationship between the developer and the construction interest, each involving competition between prospective building contractors. For example, the 'open tender', where builders respond to an open advertisement, is widely used for placing public-sector contracts, whereas commercial developers are more likely to issue an invitation to tender to a small number of selected contractors, with some of whom they may have previously forged successful business relationships. The 'negotiated price' system operates where negotiations take place between the developer and a specified contractor, or a consecutive series of such negotiations. The contract may be at a 'firm' or fixed price, although increases in price still arise from architects' variation orders and altered instructions, or it may be on the basis of a cost-plus-percentage or on an at-cost basis plus a fixed fee.

Although it is the construction process that creates the real value of any development, the model (Figure 2.5) clearly shows that construction companies are in a relatively weak position to gain access to development profits. This is because the very process of tendering for contracts obliges construction companies to compete with one another directly in order to enter into the development process at all, minimizing the extent to which developers have to surrender shares of development profit to the builders. Indeed, it has become increasingly competitive with the growing internationalization of construction companies' operations and civil engineering contractors' activities.

To some degree, the contractor also funds the project in the sense that a commitment has to be made to construction before the builder receives any payments from the developer, normally on a staged basis. Typically, interim financing of building works comprises a combination of working capital raised through short-term bank borrowings and credit advanced by suppliers and firms engaged on sub-contracts (Smyth, 1985). So, behind the main construction interest lies a network of commercial and industrial interests involved in the manufacturing and supply of building products and these relationships become reflected in complex systems of credit extended by suppliers.

Finally, it might be noted that as a major employer and consumer of building materials, it is through the construction interest that the property development process links back to the manufacturing heart of the economy. Such is the importance of the construction and development sector that it has long been used as a regulator of the wider economy in a budgetary process which became known as 'stop-go' economic policy. Development would be stimulated at times of economic slump and discouraged during booms in order to prevent 'overheating'.

2.3.5 Long-term investors (investment capital)

As property development ties up vast quantities of capital for many decades, the links between development and the investment sector are necessarily close. The creation of the built environment is therefore intimately intertwined with the operations of both short-term finance capital and longer-term investment funds because 'the availability and price of capital – in the form of mortgages, commercial loans, and direct investments – heavily condition where, when and how much commercial and residential property will be constructed at any given time and place' (Warf, 1994).

Investors comprise those interests, other than owner occupiers, which hold property rights in a completed development. Their property investment interest is undertaken in order to secure a flow of future rental income from tenants and with a view to capital appreciation (asset value growth). Indeed, 'the stock of buildings can more appropriately be seen as a stock of legal interests which serve a variety of use and investment objectives' (D'Arcy and Keogh, 1997, 691).

Commercial developers themselves frequently take on the investor role by retaining completed projects in their portfolios. For example, Olympia and York had assets worth $4 billion by the early 1980s, including 4.64 million sq. m. (50 million sq. ft.) of offices on three continents (Feagin, 1983). However, if the developer is seeking a more immediate return, the building is sold to a long-term investor. These range from small-scale individual landlords and syndicates of wealthy individuals and real-estate investment trusts, to enormous institutional investors such as pension funds, investment banks and insurance companies. For example, in 1992, the assets of UK pension funds totalled St£400 billion, 8 per cent of which was in real estate (Martin and Minns, 1995). The relative significance of the institutions has tended to vary through time and, over the last two decades, banks in Britain and North America have played an increasingly active investment role in addition to their importance as sources of short-term funds (Luithlen, 1992; Fainstein 1994).

In the commercial property investment market, developers exchange freehold rights to a finished development in return for a lump-sum payment; the market price. This price depends on the degree to which investment interests are willing to compete for available properties, bearing in mind the costs, risks and potential returns available from the alternatives to property (see later, *Yields 2.6.1*). The sale price is therefore controlled to a considerable extent by factors operating outside the property sector in investment markets generally. To investors, buildings represent just one form of investment among a range of opportunities. Property is a form of fictitious capital. It is a marketable right to the appropriation of future value that is drawn *via* the tenants of the land and building in the form of rent. This represents either a deduction from the future profits of private-sector occupiers or the expenditure costs of public-sector tenants. The major alternative investments include government securities and shares in companies. These also represent another form of appropriation of future value. In the case of company shares, the investment comprises a claim to participate in the future distribution of company profits. Government stock represents an investment in the right to receive a share of the state's future capacity to raise taxation from companies and individuals. So, all forms of investment, in their different ways, depend on the continued expenditure of future labour power (and its creation of surplus value) which can be tapped into by investors through dividends, interest-bearing loans to government, or drawn through the rental payments derived from tenanted properties.

At any time, the quantity of funds being channelled into property investment will be influenced by several factors. Most obviously, investors will compare property with other forms of investment with regard to current returns, their prospects for income growth and capital appreciation, the security of the investment and its marketability. Property meets many of these criteria well and it has been considered an appropriate investment by long-term investors for a number of reasons.

First, in an uncertain economic environment, investors seek portfolio diversification and property is an important alternative to equities and government securities. Second, it is a real asset. While shares in a company that goes into liquidation are worthless, with property, if a tenant goes bankrupt the building can simply be leased to another occupier. Third, long-term investors look for assets that will grow in value in line with inflation over the longer term. Property investment involves a high initial expenditure in acquiring a building and low expenditure thereafter, involving its management and some refurbishment on the termination of leases. Even building maintenance and insurance costs may be the responsibility of the tenant. In contrast, manufacturing involves a relatively constant expenditure on raw materials and labour power, both of which rise with inflation. So, property investment can benefit from inflation in a way that manufacturing does not. In the case of property, the initial investment is made with old (cheap) money at a large but deflating historic cost. The returns come in the form of an inflated value of rising rents. Because of this tendency to benefit from inflation, property investment has proved popular with investors seeking an inflation-proof return (in terms of both the flow of income from tenants and the building's increasing capital value) to meet fairly predictable future long-term liabilities (to pensioners and other investors).

In the UK, institutional holdings of property increased until the early 1980s, rising from St£7.2 billion in 1975 to St£27 billion in 1981, when they might typically have comprised around 20 per cent of the value of an institution's investment portfolio (Nabarro, 1990). At any point, the portfolio balance between offices, retail and industrial properties varies from one investor to another. A typical range might have been 50–60 per cent of investment by value in offices, 30–45 per cent in shops and 10–20 per cent in industrial units. While insurance companies in the UK tended to concentrate more on office property located in London, pension funds tended to have a greater involvement in industrial property (Coakley, 1994). For any investor, the precise balance between property types changes slowly according to acquisition and disposal policy, linked in turn to appraisals of their relative investment performance.

However, during the 1980s the relative significance of property investment among UK institutional investors declined as they became more cautious, property losing a certain amount of its previous attraction with the emergence of a low-inflation economy (Nabarro, 1990; Budd, 1992; Coakley, 1994; Pryke, 1994b). Although the absolute value of property held by institutional investors increased threefold between 1979 and 1991, its relative importance in investments portfolios halved. It declined from 24.1 per cent to 11.7 per cent of insurance companies' long-term investments and from 14.4 per cent to just 6.7 per cent of pension funds' assets (Coakley, 1994). With returns from UK property falling well behind those from equities and little above those of gilts, it was the banking sector which fuelled the development boom of the 1980s. It increased its lending from a negligible level in the early 1980s to having over St£30 billion in outstanding property-related loans by the end of the decade (Nabarro and Key, 1992). With banks' tendency to be more interested in the overall credit worthiness of the borrower rather than the long-term viability of particular developments, developers became freed from the inherent conservatism of the pension funds and insurance companies and undertook schemes in riskier locations (Adams, 1994). Financial deregulation and privatization also impelled building societies towards a broader banking service beyond that of mortgage lending for residential accommodation to include mortgage lending to commercial property developers.

Despite the relative reduction of property's place in UK institutional investment portfolios, the 1980s environment of deregulation was nonetheless associated with deepening links between financial markets and real estate. By late 1990, it was estimated that commercial property in private ownership in the UK totalled St£250 billion, 60 per cent of which was owned by the occupiers. Of the remaining St£100 billion worth of commercial property owned in the investment markets, some St£55 billion was owned by investment institutions (Investment Property Databank, 1991; Coakley, 1994). Similarly, Logan (1993) has suggested that the 'globalization of real estate' has encouraged a deepening relationship between the financing of real estate and the operations of wider capital markets, evident in the establishment by development companies of financial subsidiaries and their creation of long-term financial partnerships with insurance companies.

In the USA, 'commercial banks, savings and loans institutions, insurance companies, pension funds and real estate investment trusts, were loaded with money to invest during most of the 1980s' (Leitner, 1994, 793). Tempted by scarcity of supply, reflected in low vacancy rates and rising rental values, the institutions substantially increased their commercial real estate assets as a hedge against inflation and to spread risk away from the soaring, and perhaps overheating, stock market.

However, as a medium for investment, property has also traditionally posed certain problems. One difficulty is that the price of an individual property is not tested daily in the market-place as company shares would be in the stock market. The valuation of property, especially at a time of market quiescence when little comparable property is being traded, can be fraught with difficulty. Moreover, compared to the disposal of stocks and shares, property is an illiquid asset whose sale is likely to carry heavy transaction costs and may prove very difficult to sell when markets are weak (see Coakley, 1994). This makes switching the portfolio balance between gilts, shares and property difficult and leads to a degree of conservatism among investors.

Because of the high cost of buildings, each acquisition represents a sizeable commitment to a single investment item which is totally unique. This is especially problematic for investors of only moderate financial capacity. A property requires a large capital commitment for its acquisition and it may then become a sizeable component of the total property portfolio, exposing the investor to significant risk if the tenant defaults. As Ball (1998, 1504) has observed, 'buildings are a lumpy investment: they are illiquid; and, it is difficult to spread risk across a range of properties'. On the other hand, Guy and Henneberry (2000) have noted that the acquisition of a relatively small number of large, highly-priced, well-located actively-traded investment properties can also be quite attractive because it minimizes the cost of portfolio management.

These problems of lumpiness and illiquidity have been addressed over the years by a range of investment innovations, including real-estate investment trusts, property unit trusts, property bonds and the conversion of property rights into securities which can be traded on international capital markets. For instance, property unit trusts permit small investors to buy into a portfolio covering a range of properties. More recently, syndicates of investors have been assembled to purchase commercial developments. For example, in 2002, a group of 78 investors paid €82.5 million for a building in Dublin's International Financial Services Centre in units of €317,000, the balance being provided by an Irish bank on a 'non-recourse' basis limiting the investor's risk to the value of the property alone (*Irish Times*, 2002).

Many development projects, particularly the large 'groundscraper' buildings in London associated with the 'Big Bang' deregulation of financial markets, became too large and expensive even for a single institutional investor to fund. Financing required 'syndication', where a number of investors become involved as joint financiers or owners through a consortium or an investment partnership (Luithlen, 1992; Pryke, 1994a; Adams, 1994). Indeed, it was the banking sector's willingness to create such financial syndicates which facilitated the appearance of 'mega projects' beyond the lending capacity of individual banks. For example, the Broadgate office-retail complex adjacent to Liverpool Street

Figure 2.10 *Broadgate, London. This 'groundscraper' development by Rosehaugh-Stanhope, comprises 14 buildings on 12 ha. (29 acres)*

railway station on the north fringe of the City of London's financial district covered 12 ha (29.6 acres). Developed as a joint venture between British Rail and two commercial developers, Stanhope and Rosehaugh, and funded by a consortium of 21 Japanese banks, its cost had reached St£2 billion by 1991 and was still incomplete.

In Fainstein's opinion, though, the traditional difficulties have diminished:

> Previously, because of its low liquidity and unique characteristics, property investment had been the province of a limited group of financial institutions and knowledgeable individuals. Now, however, greater opportunities for real-estate investment syndication, in which limited partners did not take an active role but received an income stream and could sell their interests in the project fairly easily, eliminated any reason but rate of return to prefer one type of investment over another. (Fainstein, 1994, 30)

Finally, it might be noted that while large institutions have the financial capacity to spread investment risk across a wide property portfolio, differentiated by building type and location, smaller investors may still achieve an involvement with the property sector and achieve risk-spreading by investing in property unit trusts or purchasing shares in quoted property development and investment companies (Ball, 1998). Indeed, institutional investors also buy shares in commercial development companies and even acquire them completely, thereby obscuring their real depth of involvement with property (see Franklin, 1976; Boddy, 1981). However, this is not a true 'property' investment as, like other investments in company shares, returns depend on the managerial expertise of the commercial development company to produce profits, rather than on the rent-generating capacity based on 'property rights' in a building.

2.3.6 Tenants

It is ultimately from the users of space that the long-term returns to investors are derived. The accommodation markets provide the mechanism for the exchange of leasehold 'rights of occupation' in return for the payment of rent to the landlord, the long-term investor. Thus, the transfer of the freehold ownership from the developer to the investor will commonly take place only once tenants have signed leases, the level of rent reflecting the balance between supply and demand.

Lease provisions reflect legally the power relationship between investor and tenant. The provisions of leaseholds vary from country to country, but they will set out the period of time during which the lease will run, the periodicity of the rent reviews and the arrangements concerning the repair and maintenance of the building. For example, until very recently, the provisions contained in office leases in Ireland have greatly favoured the investor's interests over those of tenants. Leases were on the basis of a minimum of 35 years with upward-only rent reviews and the repair and maintenance costs borne by the tenant. So, at times of rent review, even if there were a glut of vacant property on the market and new leaseholds were being negotiated at lower rentals, existing tenants remain tied into paying at least the level of rent determined at the previous review. For them, rents could increase, but not decline. Only in recent years, especially at times of glut in the provision of office space when vacancy rates were high and occupiers hard to find, did tenants manage to insist on 'break options' to allow them to escape from such long leases at an earlier date.

In the UK, the typical length of office lease has shortened over the past 40 years to 15 years currently, or even less for smaller quantities of space, with tenants increasingly seeking break options. However, the frequency of rent-reviews has also increased from typically 14-year intervals in the 1960s to every 3–5 years at present. Leases of 10–15 years are common in the retail sector, while leases for industrial units commonly have, 5-, 10- or 15-year terms. In Europe, leases of 9–10 years are also common (Kivell, 1993). In contrast, in the USA, where leases can be of quite short duration, tenants have had far greater bargaining strength at times of oversupply and rental decline because they can threaten to relocate to cheaper premises.

Ultimately, because it is user demand which is being served by the property development sector, the changing geography and character of demand becomes reflected in the built environment. This has involved the large-scale suburbanization of both shopping and industrial activities since World War 2 as retailing businesses responded to the suburbanization of consumer demand, while industrial companies were encouraged by operational considerations and planning authorities to relocate to less congested peripheral sites. Office functions followed, often in response to escalating central-city rents and encouraged by improvements in transport infrastructure enhancing the accessibility of the urban edge.

The quality of buildings also improved, particularly of office premises and industrial property. This became reflected in the almost obligatory provision in offices of features such as air conditioning systems, high-performance lighting and raised floors to accommodate IT communications systems. With suburbanization came the development of well-landscaped office-parks; a product often aimed at attracting international blue-chip corporate occupiers.

For industrial premises, change is even more apparent. The speculatively-developed industrial sheds of the 1960s, densely developed with little parking or provision for deliveries, often using concrete blocks and asbestos roofing, commonly provided little office accommodation. These gave way to an altogether higher quality product as user demand changed. Business operations paid increasing attention to the storage volume of buildings rather than merely to floorspace, developers responding by raising eaves clearances and by increasing the weight-bearing capacity of flooring to permit higher stacking. The range of construction materials also widened to include metal deck roofing, high-technology glazing and profiled, insulated, coloured metal cladding for walls. More generous provision of space for deliveries and landscaped car parking resulted in a reduced development density, with buildings covering as little as 30–40 per cent of the site.

Businesses that might hitherto have conducted managerial operations from central-city offices, separated from their suburban industrial operations, increasingly sought to amalgamate operations at the periphery. The quality of office accommodation required therefore rose, marked by the appearance of atrium reception areas, suspended ceilings, modular light fittings, carpeting and raised floors for computer cabling. The proportion of the building comprising the office component also increased, perhaps accounting for over half the floorspace of buildings accommodating high-technology users (see Henneberry, 1994).

Thus, there has developed a tendency towards a blurring of the functional categories of accommodation. The new forms of space maximize in-built flexibility by providing equally for light-industrial or office uses, becoming reflected in the UK zoning category B1. In large part, this flexibility reflected changes in user demand as new types of industrial/office activities appeared, seeking out 'science park' locations for the production of high value components, computer software development, for research functions or 'call-centre' operations. But it also resulted from developers' own desires to minimize development risk by maximizing the range of potential users. Such properties are therefore set somewhat apart from those of normal industrial operations, typified by a greater rental capacity of users and enhanced potential for rental growth. These factors, together with the higher quality of construction and lower rates of obsolescence and depreciation become reflected in different profiles of investment performance.

2.4 Considerations

Several points are worth emphasizing with regard to the model.

1. Clearly, new developments result in response to the potential for property development to act as a vehicle for generating profits for those forms of capital involved in the creation or redevelopment of the built environment. Although the nature of the returns to each of the interests is rather different, these are all ultimately derived from those who occupy the building. Thus, if no users are forthcoming, no profits will accrue.

2. Clearly too, the archetypal interests involved in the property sector are heterogeneous and the interaction between them is mediated through distinct markets, which may be structured more or less formally. Consequently, as property development requires the profitability criteria of each participating interest to be fulfilled, the production of the built environment cannot be treated as an unproblematic reflex response to changing conditions of demand in the accommodation markets alone. Although the accommodation markets reflect the strength of demand for buildings, thereby linking the development process to the cyclical trends in the wider economy, the fact that these conditions are mediated through the relations of production within the property sector means that the response becomes unpredictable.

For many involved in the industry, engagement in the property sector is but one of a number of possible options for pursuing capital accumulation. If at any point in time the terms of engagement of any one of the interests is not met, development is unlikely to happen. For example, investors might believe that equity shares in companies offer more lucrative opportunities than property, or perceive government securities to provide a less risky investment. They can put their money into alternatives to property, or invest in other property markets worldwide. The absence of a long-term 'take-out' represented by prospective investors inhibits further development.

3. Although many of the personnel from the various interests operating in the property development sector become well known to one another, creating a web of well-established working relationships which engender trust, the relationships between the archteypal interests exist in a state of dynamic tension. This is because of the inherent competition between the interests over the distribution of development profits, the differences in their economic power and the varied duration of their engagement.

4. It can also be seen from the model that the property development industry serves two markets: the accommodation market servicing user-demand and the property investment market serving investors requiring buildings for their flow of rents. These two markets can get out of synchronization, the operations of one not always equating with the functioning of the other, as will be seen later when reviewing development cycles. At times of boom, the rush of investors seeking to channel funds into property may drive valuations to unrealistically high levels and encourage developers to initiate too many schemes, aiming to meet demand from investors rather than from users. Inevitably, the market tends to 'overheat' and many more buildings are developed than user-demand justifies.

In an era of increasing globalization, there is heightened risk from intense overheating as large-scale international flows of capital move into property (Ball, 1994; Coakley, 1994). For example, continual Federal budget deficits and borrowings during the 1980s in the USA, resulted in high real interest rates attracting huge inflows of capital from Japan, Germany and Canada, much of which found its way into real estate. By the 1990s, half the commercial property in downtown Minneapolis was owned by Canadian corporations and half that of downtown Los Angeles was Japanese-owned (Warf, 1994).

5. Financial institutions have often secured their involvement with property by forward funding schemes at development stage. At times, this has almost relegated the role of the developer to that of project manager. Although this strategy reduces the developer's profit while increasing that of the institution, advantages do also accrue to the developer by providing a guaranteed 'take-out' for the scheme, thereby reducing risk. Institutions also undertake property development on a direct basis, permitting them to capture the development profit entirely and giving them greater control over the quality of schemes.

6. Finally, with regard to the theorization of urban development, Harvey (1981, 1982) has argued that investment switches into built environment creation (the 'secondary circuit' of capital) when 'over-accumulation' occurs in the 'primary' (productive) circuit. However, Fainstein (1994) observes that the distinction between these two circuits has reduced as the changing character of property lending, notably syndication and securitization, have made property debt increasingly negotiable (see also Ratcliffe and Stubbs, 1996).

Moreover, although depressed conditions in the wider economy may limit competition for capital and reduce the cost of borrowings, this is merely one positive element in calculating development viability. The model shows that the returns to all the interests ultimately flow from the users of the new building. Completed development must therefore be capable of attracting occupiers. But just as you do not buy an envelope unless you want to post a letter, so businesses do not buy or rent buildings unless they need to accommodate new or expanded operations. Neither do developers normally build with the intention of creating vacant space. As the scale and character of user demand for accommodation is ultimately closely tied to conditions within the wider economy (the 'primary' circuit), the concept of property development providing a convenient dumping ground for capital is counter-intuitive. Ease of short-term borrowing is a positive factor in the equation, but it is not a sufficient condition for development to take place.

Certainly, developments undertaken during slumps may benefit from lower construction costs, cheaper financing and possibly lower site acquisition costs. However, even the large speculative development of the Empire State Building (completed in May 1931) seems to have been taken *inspite* of the Wall Street crash of October 1929 rather than in order to reap the benefits of slump conditions. Deciding to continue with construction at least offered some prospect of a return on the massive investment already made, whereas cancellation would have meant inevitable loss. Nevertheless,

Figure 2.11 *The Empire State Building, New York, completed in 1931*

much floorspace remained vacant until World War 2, some suggesting that it might be renamed the 'Empty State Building' (Knevitt, 1985). In contrast, the development of the 18-building Rockefeller Center did result from the developer's taking a well-considered long-term perspective and the benefits to be derived from lower development costs.

At times of slump and depressed real-estate prices, there is also very little empirical evidence of investors being prepared to purchase buildings which, due to weak user demand, are unable to secure tenants to provide the required rental flow (see MacLaran et al., 1987). Similarly, there is a reluctance of banks to finance their development. For example, following 10 years of continuous economic growth and the greatest office development boom in Dublin's history, the economic downturn in 2001 saw major over-provision of space in suburban office parks. At one 48,000 sq. m. (516,000 sq. ft.) office campus, vacancy reached 70 per cent by June 2002. In the course of a single year, quoted rentals there fell from Ir£160 per sq. m. (Ir£14.80 per sq. ft.) to under Irf90 (Ir£8.40). Development finance was withdrawn, long-term investment funding evaporated and construction was suspended.

2.5 A profitable business?

The driving force for the property development is the search for profits and, without doubt, it can be enormously profitable. In the UK, urban reconstruction after the destruction of World War 2, the reconfiguration of city centres through the 1950s and 1960s, increasing office-based employment, growing demand for new retailing and industrial buildings, together with the accommodation requirements of an expanding population, created major opportunities for developers. Oldham Estate, which was taken over as a valueless 'shell' company in 1959, was built under the direction of Harry Hyams into an operation valued at over St£100 million by 1967 (Marriot, 1967), rising to around St£300 million prior to the property crash of 1973/4. The rise of Harold Samuel's Land Securities was even more spectacular. Between 1944 and 1985, its asset valuation rose from less than St£20,000 to some St£2,335 million (Green, 1986), topping St£3,675 million by 1993.

Profit can be derived in a number of ways. First, a developer can decide to take a 'developer's profit' and sell a completed building to an owner occupier or to a long-term

Figure 2.12 *Centre Point, London. A development by Harry Hyams' Oldham Estate Company*

investor. The developer profits insofar as the capital sum received exceeds the total cost of undertaking the development. Alternatively, the developer can retain the ownership of a building and receive rental income from tenants. Effectively, in this scenario, the developer takes on the role of investor by placing the property in the company's investment portfolio. This is profitable insofar as the rental income that the property generates exceeds all the outgoings. These comprise mainly the interest charges on borrowings to finance site acquisition, construction and professional fees, together with any repayment of the principal, although loans are frequently advanced on an 'interest only' basis for the first five years. However, it is unlikely that the level of initial rental income received from the first letting of a property will cover its costs. In the initial years, rents are set in a competitive market. Expensive new buildings have to compete against older vacant stock coming back onto the market as leases terminate or as tenants' businesses fail. New property built on high-cost land in the office core also faces competition from other locations, including cheaper suburban green-field sites and inner-area redevelopments, such as former docklands, where land may be made available at a subsidized price by a public redevelopment authority. Cheaper development costs can then be reflected in very competitive rental levels.

Large, well-established commercial developer-investors, such as MEPC in the UK which had an annual rental income of St£320 million in 1992, have built up an international property portfolio yielding a substantial rental flow. They can use the 'surplus' income from existing investments to cover the shortfall in income from a new development. But for the small commercial developer, the involvement is likely to be short-term with the building being sold on to a long-term investor. The course chosen, whether to dispose or retain, will reflect a number of considerations. They include the predicted future performance of the property in relation to the level and security of rental income, the prospects of capital growth and the size of the capital investment represented by the building in relation to the remainder of the investment portfolio. For example, an expensive office scheme might represent an overly large exposure in one market and create an imbalance between the value invested in offices, shops and industrial buildings in the portfolio.

For institutional developers, such as the property development arms of life assurance companies, it may be possible to adopt a longer-term investment perspective, particularly in the past. Rental flows comprise just one element of total income, including life insurance premiums or pension contributions that provide a continuous flow of new investment funds. However, in recent decades, especially with the expansion of single-premium investment business during the 1980s, growing competition for savings funds heightened the expectation among investors of a more immediate performance by the institutional investor-developer. It therefore became increasingly difficult to justify investing in property as it failed to perform competitively over the shorter term in a low-inflation highly-competitive investment environment (see Pryke, 1994b).

For the cash-strapped commercial developer, there exists the possibility of a sale-and-leaseback arrangement. Here the developer can receive an immediate capital sum by selling the freehold of a property on to an investor (the sale), who then grants the developer leasehold rights to the building (the lease-back). The developer subsequently grants sub-leases to tenants. This has the advantage of providing an immediate lump sum return to the developer, thereby raising funds which can be used for further development or to amortize the debt which has been incurred in undertaking the development. It reduces the scale of outgoings and shields the developer from the potential disaster which any rise in interest rates might occasion for a highly geared (debt-laden) company. Yet, during the period of the 'lease-back', it still permits the developer to benefit from future increases in rental levels. Unlike a mortgage loan, it also permits the institutional investor to share in the rising rental and capital values of the property (see Adams, 1994).

Thus, for any particular scheme, the developer's objective will be influenced according to whether it is being carried out by a small-scale entrepreneur, by a large commercial developer or by the development arm of an investment institution. Small-scale operators are more likely to look for immediate capital gains and are more likely to operate by acquiring a site, developing it and selling the completed and tenanted building to a long-term investor for a capital gain.

Of course, property development does not always prove profitable. If development costs spiral out of control or if the project reaches completion at the wrong time, during a slump in the demand for accommodation, it can mean liquidation for the commercial developer. Thus, while risk minimization is important to all developers, it is absolutely crucial to the small-scale commercial operator. At any stage in the process, weighing risk against the prospects of higher development profits will be a key determinant of success or failure. This tests the developer's ability correctly to read the market. For example, following site acquisition, a number of choices have to be made. The demand for accommodation generally may be great enough to create sufficient confidence to develop speculatively without a specific tenant or purchaser to hand. Alternatively, construction may be delayed until the prospective development has been 'pre-let' or 'pre-sold'. Obviously, speculative development is less risky in certain property sectors, such as retailing, where the boom-slump cycle tends to be less marked.

Further risk reduction might be effected by pre-selling the development to an institution after pre-letting but prior to construction. Alternatively, at an even earlier stage in the process, the developer may enter into a profit-sharing arrangement with a financial institution whereby the developer agrees to sell the site to the investor, who then funds construction. The use of such forward funding may extend to a full buy-out guarantee, in which case the developer's role almost becomes relegated to that of a project manager, as was noted earlier. Indeed, at times of credit scarcity, institutions may actively use their financial power to buy into developments. These arrangements reduce the developer's share of any profit, but they also reduce risk. A long-term 'take out' is secured, a profit is generated from the sale of the site and a share, albeit lower, in the development profits is guaranteed. Moreover, it eliminates risk of exposure to rising interest rates.

2.6 The driving force: profit

We saw above that the catalyst for speculative private-sector property development is the prospect of the profit to be realized from development and that the key to success is that the market value of a completed building should exceed its cost of development. In calculating development viability, the major cost elements must be evaluated (see Cadman et al., 1991). These include the price of the site, which may already be known or may yet to be determined, the estimated construction costs, the cost of professional and consultancy fees and the interest charges on borrowings. The calculation of prospective development profitability therefore has to be made in a climate of considerable uncertainty (Ball, 1994). Development appraisal is rendered more complicated by the fact that even within a single sector, land and property markets are increasingly segmented and highly differentiated (Healey, 1998). For example, late in 2001, vacancy rates in modern office buildings in Dublin ranged from under 4 per cent in the prime office core to over 34 per cent in western suburban office parks.

The key element in the calculation is, of course, the price achievable for the finished building. Briefly, this is determined by the rental income it generates and the acceptable level of initial yield which this represents to an investor, according to the formula:

$$\frac{rental\ income}{yield\ \%} = Capital\ value\ (price) \qquad e.g. \qquad \frac{€100,000}{5\%} - €2,000,000$$

Thus, a building costing €2 million renting at €100,000 generates an initial yield of 5 per cent for the investor. Clearly, as rents double, capital values also double. However, if the prospects for future rental growth look good, more investment interest will be shown in property and investors compete more determinedly for available properties. The effect of growing competition between investors for a relatively fixed stock of investment property is to bid up its price. The price that the investor has to pay in order to obtain that flow of rental income increases. Thus, the initial yield which this rental income now represents falls. So, as capital values rise, initial yields fall or 'harden', perhaps to as little as 4 per cent:

$$\frac{€100,000}{4\%} = €2,500,000$$

So, the sale price of a building may bear little relationship to its cost of construction. Indeed, Barras (1979a) suggested that during booms in the City of London office market, when rents were high and investors were willing to accept a very low initial yield, construction costs may comprise only 10 per cent of the price. This might rise to around 30 per cent during recession. For provincial cities, the estimated construction element ranged between 40 and 70 per cent. Clearly, if the timing is correct, property development can generate an enormous pool of profits for which the private-sector interests compete. If incorrect, the level of anticipated rent, upon which the profitability of the scheme was calculated, may not be achieved. Worse still, the building may remain vacant.

Thus, the capital value of a building is established by reference to the two separate markets which the industry serves. Prevailing rents are set in the accommodation markets while the level of acceptable initial yields are established in the property investment markets. It is therefore important to investigate these two elements further.

2.6.1 Yields

The level of initial yield deemed acceptable from property is set in relation to the returns available from other investments, their liquidity, their future prospects and their levels of risk. Surprisingly, initial yields from prime property may be several percentage points below risk-free government stock. This is because investment in property is made in the hope that rents will rise with periodic rent reviews and ultimately outstrip the returns from gilts (see Figure 2.13).

However, real estate is far from being a homogeneous form of investment. The retail, office and industrial (factories and warehousing) property sectors perform rather differently as investments. Each has a different prospect for rental increase and capital growth, liability to obsolescence and rental default. For any one sector, this is further complicated by variations in the age of buildings, their construction quality and location, high quality buildings in 'desirable locations' that are in strong demand by tenants being typified by lower levels of acceptable initial yield (i.e. larger multipliers of the rent). To explore this, the factors influencing yields will be outlined for the retailing, offices and industrial property sectors.

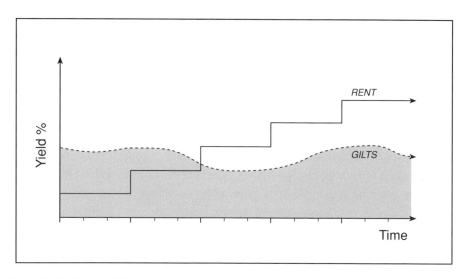

Figure 2.13 *Erosion of the yield gap between gilts and property resulting from rent reviews*

2.6.1.1 Investment considerations (i): retailing premises

For retailing properties, an initial investment choice must be made between the city centre and suburbia. However, even within the central city, units on prime shopping streets will offer different investment opportunities compared to secondary streets. Modern central-city shops in prime locations comprise uncomplex buildings built on expensive land, highly accessible to potential customers. The supply of central-city prime retail sites is also limited and competition between retailers wanting to locate there is intense. If the street has not become subject to 'locational obsolescence', where prime retailing has moved elsewhere, even if the current tenant ceases to trade, the prospects of vacancy (a 'lettings void') is low. When leases terminate and buildings require replacement, the site has a high 'residual value'. The long-term investment security of such a location with good prospects for rental growth may induce an investor to accept an initial yield between 4–7 per cent on the property's purchase price. In fact, initial yields on prime central-area retail properties are amongst the lowest to be found in the property investment market.

Secondary streets, on the other hand, offer lower entry costs for the investor and the prospect that general growth in retailing may generate upgrading of the secondary location to incorporate it into the core. It might be riskier, but there may also be the prospect for spectacular rental increase and capital growth if overspill occurs. While suburban retailing is more likely to face competition from the development of neighbouring malls, vigorous residential suburbanization has generated enormous local custom with housing development booms fuelling the demand for traditional out-of-town retail warehousing products (e.g. Do-It-Yourself retail centres, garden centres, furnishings, etc.). Indeed, the out-of-town retail element is typified in 2002 by amongst the lowest initial yields in the UK property sector, reflecting its strength and prospects for rental growth.

At a micro-scale, the preferences of retail tenants become reflected in the appraisal of potential investment properties. Different ends of the same shopping street may have contrasting prospects for investment performance. One end may be enhanced by the opening of a new 'people draw', such as a prestigious department store or an in-town shopping mall, while a degree of locational obsolescence may simultaneously creep in elsewhere.

The success of retailing businesses depends on their ability to lure passing pedestrians into shops to make a purchase. This determines the level of rent which tenants can afford. The volume of pedestrian flows are therefore crucial. Corner sites facing two streets offering window displays on two sides have an advantage over mid-row locations. In broad terms, on prime retailing streets, rent might equate to around 5–8 per cent of retail turnover. But not all shop floorspace is of equal value to a retailer. The ground floor is more valuable than basements or upper levels. Even the front of shops has greater value than the back and is reflected in the 'zoning' of space within individual units (in depth bands of 20 ft./6.096 m. in the UK). The most expensive Zone 'A' rents are charged for space at the front, with progressive reductions as depth increases. Thus, for a given floor area, units with large frontage and shallow depth will command higher rent.

2.6.1.2 Investment considerations (ii): offices

Location is a key element for investment consideration, decisions having to be made between the merits of prime city-centre offices, neighbouring secondary sites and suburban locations. Building quality and the date of construction are obviously also important factors, converted nineteenth-century buildings having a different investment profile to modern purpose-built stock with air conditioning, raised floors and ample car parking.

Well-constructed 'state-of-the-art' premises located in the central city, where sites are in limited supply and highly sought generally possess a good residual value. After 20–25 years, with refurbishment and modernization, older buildings in prime locations may command a rent little below that demanded for new premises. The prospect of long-term vacancy is low due to the high user-demand for the limited supply of such prestigious locations and even lower if a public-sector tenant is secured.

In spite of technological advances, particularly in communications technology which permit tele-conferencing, the inertia of business operations and other considerations such as the opportunities for personal and career-related 'networking' of staff, or simply the 'buzz' of central-city business life, continue to ensure a thriving office core in most cities. With positive long-term prospects, the level of initial yield demanded by investors for prime modern offices is therefore low, typically ranging between 5 and 8 per cent. However, secondary offices are often badly affected at times of recession and the higher yields demanded reflect this tendency.

2.6.1.3 Investment considerations (iii): industrial buildings

In a sector where owner occupation of custom-built premises had been traditional, the use of industrial properties as a medium for investment commenced relatively recently (Fothergill et al., 1987). However, the development of standardized industrial units appropriate to various manufacturing or warehousing functions created a tradable asset suitable for long-term investment (see Ball and Pratt, 1994). Additionally, investment-quality properties were often available at far lower unit costs than for offices buildings, thereby facilitating the entry of small-scale investors into the property sector.

Investment appraisal will take account of the historically modest rental growth of industrial units and their propensity to lettings voids. They may also be subject to rapid depreciation through wear and tear and general obsolescence, reflected in their long-term rental performance. A 20-year-old industrial building is probably of significantly poorer structural quality and amenity provision compared to a new building, commanding a considerably lower rent. The residual value of industrial units is also generally low, reflecting the low land value element. Indeed, the cost of breaking up the concrete base and dealing with materials such as asbestos, previously commonly used, means that green-field development on virgin land may well prove cheaper than redevelopment. Consequently, the original investment has to be recouped over a much shorter period compared to shops or offices. Thus, the level of initial yield required from industrial property is commonly far higher, ranging between 7 and 11 per cent.

However, the character of industrial space has changed considerably over the past 20 years. Recently-developed industrial premises have different construction materials, high-quality associated office space and landscaped car parking. They give the appearance of an altogether higher level of attention to design and quality of development compared to the past, more akin to office parks than industrial estates. This evolution has resulted in changing investment appraisals and the emergence of a recognizably distinct subsector of the industrial property market (Henneberry, 1994), typified by lower initially required yields.

2.6.2 Rents

The second element of the price equation is rent. Rent is the economic reflection of the legally-enshrined social relationship of 'property right' in which rental payments are made for the right to occupy land and/or buildings The scale of rental payments which can be extracted from a plot or a building is determined by four elements: the class of user, the type of function accommodated, the quantity of space generating rental income and the general conditions of supply and demand in the accommodation markets. Of these four, only the latter lies beyond the influence of the property owner.

Clearly, from the equation above (p. 37), if rents double and yields remain constant, the value of the building doubles and the land on which it is built also increases in value. Because land price (capitalized ground rent) is determined in relation to the revenue-generating capacity of the site, there exists a constant pressure on property owners to upgrade land use in order to increase the flow of rental income. This is undertaken in three main ways.

1. Upgrading the class of the occupier. Typically, this might lead to low class locally-owned shops being displaced by international multiples or franchises which possess a greater drawing power and rent-generating capacity per unit of floorspace. This process of displacement is akin to 'gentrification' in housing markets, though without the transfer of tenure which usually characterizes the change from working-class tenants to middle-class owner-occupiers. However, businesses in owner-occupied premises may continue to engage in less profitable activities and resist upgrading because freeholds provide a degree of insulation from pressure to upgrade business operations compared to being faced with ever-rising rents (D'Arcy and Keogh, 1997).

2. A second method of upgrading is to change the type of function accommodated. This is the basis for converting agricultural land to urban functions. Similarly, in urban areas, low-grade functions are displaced by more remunerative uses, leading to the adaptive re-use of buildings (see Figures 2.14 to 2.19) and to the redevelopment of sites for higher-grade functions.

Figure 2.14 *Ghirardelli Square, San Francisco. Adaptive re-use of a chocolate factory accommodating cafés, restaurants, a theatre and 75 shops*

Figure 2.15 *Covent Garden, London. Retailing in London's former fruit and vegetable market*

Retailing and offices normally have a higher rental capacity than either industrial or residential uses. Despite the growth of Internet shopping for goods and services, there remains in both retailing and office-based services a measure of locational monopoly which can be reflected in the ability of retail and office functions to pay higher rents for desirable locations. In contrast, manufacturing industries produce physical commodities that commonly compete on world markets and the element of spatial monopoly is likely to be limited. Simultaneously, the current technology of industrial production and warehousing favours expansive (land-using) single-storey buildings, too expensive to accommodate in central areas where land prices are high. Such differences in the rental capacity of various types of user therefore lead to the displacement of residential and industrial buildings by commercial functions (offices and shops) and to the conversion of industrial sites and warehouse buildings to luxury residential use. It is therefore not uncommon to discover a 'softening-up' process underway at the edge of the CBD as

Figure 2.16 *Stamford, Lincolnshire, UK. With declining church attendance in the UK, religious buildings are put to new uses*

Figure 2.17 *Butler Square, Minneapolis. A warehouse (built 1907) converted to offices and retailing*

higher-value uses start to move in (Figure 2.20). However, this may well encounter opposition and lead to conflict with those being displaced (see Figure 2.21 and 2.22).

As industrial operations seek to cash in on rising site values by selling out and relocating, their displacement may contribute to an 'inner-city crisis' resulting from decling industrial employment opportunities suited to the skills of inner-city residents. Just such a process of 'upgrading' took place in New York. Development forces acting in concert with

Figure 2.18 *Butler Square, Minneapolis (interior). Two atriums open up the building's volume, shown to good effect by George Segal's 'Acrobats'. Heavy structural timber has been sand-blasted and left exposed*

Figure 2.19 *Gun Wharfs, London docklands. Conversion of warehousing for residential functions*

planners underlay a wholesale 'de-industrialization' of New York's economy where, as late as 1922, almost 420,000 factory workers were located on Manhattan Island south of 59th Street, which had preserved a diversified economic base into the 1950s (Fitch, 1993). These were displaced by the office-based finance, insurance and real-estate sectors, creating a ten-fold increase in the returns to property owners compared to industrial buildings.

However, changing from one type of function to another may require significant restructuring to the fabric and layout of a building. For this reason, urban planning systems may 'list' for protection the facades and interiors of buildings to prevent their being externally modified or gutted internally and inappropriately reconfigured.

3. Augmenting rental income from a site can also involve increasing the quantity of rentable space. The volume of an existing building might be reconfigured to reduce the space devoted to stairways and halls that are unproductive of rent. Alternatively, a developer might replace a building with a more space-efficient structure, perhaps simultaneously also seeking to increase the plot ratio by constructing a much larger building (see Figures 2.23 and 2.24). Whereas the earliest 'skyscrapers' had load-bearing external walls measuring some 2 m. (6 ft.) in thickness at ground-floor level, the invention of the 'curtain wall' iron-frame and, later, steel-frame building permitted the 'fight for

Figure 2.20 *The Rocks, Sydney. Adjacent to the office core, the district was long under threat of encroachment by commercial redevelopment*

height' to commence in earnest. It is a process which has driven the skyline of cities worldwide to become dominated by skyscrapers (Figures 2.25 and 2.26)

Again, planning authorities commonly intervene to limit the impact on urban environments of these pressures to redevelop sites to ever-greater densities, placing preservation orders on buildings and, in the case of permitted redevelopment, by controlling building heights and plot ratios.

It is the potential profit from high-value redevelopment which underlies the high cost of central city land. But equally, it is the high cost of land that determines what can be developed there at a profit. Thus, the urban land use patterns which emerge are in large measure an outcome of market-led decision making. To developers, the urban landscape represents a profitability surface that is highly differentiated geographically. The arrangement of this surface is such that market forces may direct attention away from

Figure 2.21 *Redevelopment of Tolmers Square, London*

Figure 2.22 *Redevelopment frequently encounters opposition, as at Tolmers Square (see Wates, 1976)*

Figure 2.23 *Bligh Street, Sydney*

Figure 2.24 *Bligh Street, Sydney. Successive land-use intensification in the twentieth century has transformed Sydney's commercial core*

areas where renewal is required towards those in which potential rental values are at their highest and where initial investment yields are at their lowest. These are commonly located in the very districts where the urban fabric is at its highest quality.

For example, to buy a 10,000 sq. m. building renting at €200 per sq. m. in the prime area of the city, an investor might be willing to accept an initial yield of 5 per cent, believing that rental growth will be strong. This equates to a building price of €40 million.

Figure 2.25 *Downtown San Francisco*

Figure 2.26 *Massing of skyscrapers, Manhattan*

A similarly sized building in a secondary location might generate only €120 per sq. m. Moreover, the initial yield required by investors would typically also be higher, perhaps demanding 8 per cent to reflect the riskier location. The effect on the development's sale price, reduced to €15 million, and its potential profitability is enormous. Although land costs might be lower in the second location, construction costs would be similar and short-term development finance may even be more costly, reflecting the riskier nature of the development.

This understandable 'preference' for prestige locations and the avoidance of others where investment and renewal is needed has frequently led governments and planning authorities to provide financial inducements in order to alter the market-based profitability surface in the hope of shifting the geographical focus of development interest. It is an approach that is taken up in the Dublin case-study.

2.7 Property cycles

Property development is subject to boom–slump cycles and 'as in all crises of production, the anarchistic element of overproduction is at the heart of property cycles' (Luithlen, 1993, 29). Such cycles have been identified in virtually all the major sectors, but they are particularly evident in the office, industrial and residential markets (see Barras, 1979b, 1983; MacLaran et al., 1987; Leitner, 1994; Moricz and Murphy, 1997; Dehesh and Pugh, 1999, 2000). Figure 2.27 depicts annual office development in Dublin over a period of more than 40 years. The impacts of fiscal incentives to property development in the late 1980s and unprecedented economic growth of the 1990s, the so-called 'Celtic Tiger', are clearly evident in the third and fourth booms respectively. In just two years, 1990 and 1991, the stock of modern office space expanded by 18 per cent, while the post-1995 boom generated an 80 per cent increase in the modern stock.

Numerous factors have been cited to account for development cycles. Some factors are clearly exogenous to the property sector, such as the availability and cost of finance and the relative yields from competing investments (see Barras and Ferguson, 1987). Ultimately, the scale of development activity in all property sectors relates to the level of user demand for buildings and, therefore, to the more general performance of the economy. Nevertheless, the different property sectors may be in somewhat different phases. The industrial sector might be in a phase of quiescence while retailing may be quite buoyant. However, if the booms or slumps do coincide, the consequences for some participants in the development industry, particularly the construction sector, can be very severe. Other factors are endogenous, deriving from the inherently lengthy period required for development: from preparatory phases, through construction to completion. A good discussion of building cycles is provided by Leitner (1994).

2.7.1 Upturn and acceleration

The stimulus to development activity is the demand for buildings to accommodate particular functions. In his examination of the almost worldwide property development boom of the 1980s, Ball (1994) cited broad economic processes as the driving force, showing how technical change and de-industrialization acted as catalysts. Shifting patterns of demand in key service industries generated an expanded need for modernized and new buildings, particularly in 'world cities' such as New York and London (see Ball, 1994; Coakley, 1994; Fainstein, 1994). Demand for accommodation from 'high-tech' industry, 'big-box' retailing and the deregulated financial markets underpinned the property boom of the 1980s. It often involved larger units built to a higher quality and in completely new locations. The resultant shifts in patterns of employment in turn created an upsurge in regional housing markets.

Whatever the initial demand-based stimulus, buildings take a long time to produce, particularly large commercial developments. The stock of floorspace therefore remains in relatively fixed supply. Under market conditions where increased demand for a product is met by inelasticity of supply, shortages result and prices increase. In the case of buildings,

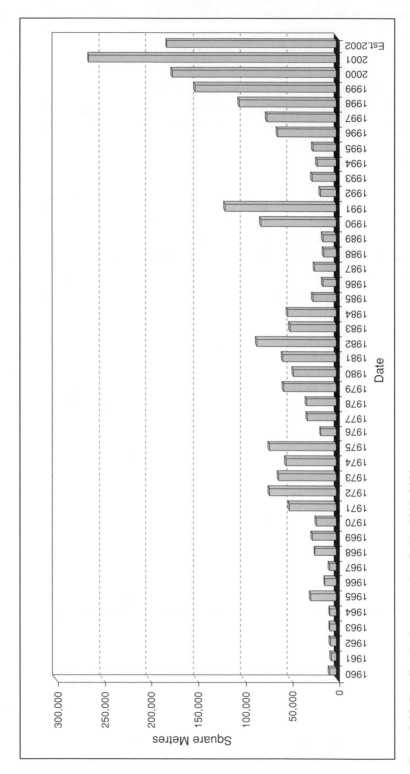

Figure 2.27 *The office development cycle in Dublin, 1960–2002*

increased demand generates upward pressure on rents and therefore on capital values. As rents rise, developers with ready-to-go schemes awaiting signs of increasing demand, start building. Others are encouraged to hasten site assembly and gain permission from planning authorities to embark on new developments. However, development is a slow process, even for projects where construction can start immediately. So it takes a considerable time until the new schemes can meet the rising demand from prospective users. Thus, there exists an extended period when rents and prices continue to rise.

2.7.2 Boom

Escalating prices and the prospects of high profits encourage increasing numbers of developers to initiate developments. In addition, as Ball (1994, 682) has noted 'property booms are always associated with relative newcomers to the development world. Generally, these entrants start with little own capital and borrow to undertake large-scale developments. Even during upturns many entrants will not be successful; but some will'. Local developers are joined by outsiders, commercial developers constantly reviewing opportunities internationally. Thus, Canadian developers came to control a quarter of all US office construction in the period 1979–82 (Feagin, 1983, 68). In periods of boom and high potential profitability, the geographical field of development activity also commonly tends to widen, with office developments often being undertaken in locations which would have been considered far too risky at more sober times (Malone, 1985; MacLaran, 1993; Pryke, 1994a).

2.7.3 Investment-driven development

With the prospect of rents and capital values rising into the future, property looks like a good vehicle for investment. The bulk of decisions taken regarding commercial property investment are made by relatively few portfolio managers (Martin and Minns, 1995, 135) and a certain 'herd instinct' can be identified among them. Those recognizing the potential of property investment early in the cycle do well and are followed by other investment managers, fearful of neglecting such an obvious investment option. Investors, each responding to the same stimuli and market awareness, vie to outbid one another for available buildings. International flows of investment capital also target property markets exhibiting strong performance, thereby adding to the pool of funds seeking investment outlet (Thrift, 1986). In their enthusiasm for investing in property, building price is bid up, forcing down the level of initial yield often to absurdly low levels. The price of a building renting at €100,000 per annum might therefore be bid up from €2 million (representing a yield of 5 per cent) to €2.5 million (a 4 per cent yield) trusting that rentals will rise in the future.

The upward bidding of building price by the investment market in turn encourages more development activity than a careful reading of the accommodation market would justify. However, no mechanism exists to ensure that the production of buildings will be exactly tailored to match user demand. Thus, there exists an inherent tendency towards the over-provision of space. Developers start building with a view to satisfying the investment demand for buildings, rather than for the real level of user demand. As increasing amounts of investment-driven development are undertaken, the market overheats and far more floorspace is added than user-markets alone could justify. During the development boom of the 1980s, London's office stock increased by 4.15 million sq. m. (44.7 million sq. ft.) to 14.37 million sq. m. (154.7 million sq. ft.), a large proportion of which was the product of just five developers: Olympia and York, Speyhawk, Stanhope, Greycoat and Rosehaugh. Similarly, New York's stock rose by 4.9 million sq. m. (53 million sq. ft.) to 22.57 million sq. m. (243 million sq. ft), with 1981–2 alone seeing the completion of 12 new office towers totalling over 650,000 sq. m. (7 million sq. ft.) (Fainstein, 1994).

Even when it becomes apparent that not all the new developments can secure occupiers, individual developers tend to take a positive view of their own schemes, regarding them in a more positive light than competitors' schemes. In any case, the development process cannot be suspended without significant cost. Although a building may be under construction when signs of impending downturn first appear, it either has to be completed despite the diminishing prospects of attracting an occupier, thereby throwing good money after bad, or the considerable investment already tied up in the half-finished building has to be written off. Property development is like steering a super-tanker at sea; it is impossible suddenly to slam on the brakes and stop.

2.7.4 Over-supply

With growing volumes of development reaching completion, developers find it increasingly difficult to attract tenants. Unforeseen costs are also incurred. More has to be spent on the development, perhaps completely fitting out the building to attract an occupier, or offering prospective tenants inducements such as rent-free 'holidays'. Inevitably, as user-demand is satisfied, the pressure for rental increase is eased. However, there may still remain significant quantities of floor space in the process of construction. An over-supply occurs and vacancy rates rise rapidly.

Nonetheless, office rents may respond only slowly to falling demand, continuing to generate inflated rates of return for property and encouraging yet further development in spite of rising vacancy (Wheaton, 1987). Eventually, however, rents inevitably stabilize and even fall. Thus, in the USA, the 'overhang' of vacant space which caused national office vacancy rates to quadruple from 5 per cent in 1980 to 18.9 per cent in 1991 and which pessimists considered might take 14 years to absorb, became reflected in real average rental rates dropping from $172 per sq. m. ($16 per sq. ft.) in 1980 to less than $107 per sq. m. ($10 per sq. ft.) in 1990 (Warf, 1994; Leitner, 1994). With 6 million sq. m. (65 million sq. ft.) of vacant office space in New York in the early 1990s, equivalent to 50 years of demand, at a time when office development costs were $3,230–4,300 per sq. m. ($300–400 per sq. ft.), one office building off Wall Street was sold in 1993 for a *price* of just $118 per sq. m. ($11 per sq. ft.). This was less than the prevailing *annual rental* value five years earlier (Fitch, 1993). Similarly, the aftermath of the 1980s London office boom created a legacy of 1.58 million sq. m. (17 million sq. ft.) of vacant space by late 1991 and prime rents 35 per cent below their 1988 peak (Pryke, 1994a).

2.7.5 Bust

With growing vacancy, capital values drop as investors recognize that there is limited likelihood of rental growth in the near future. They therefore draw back from further acquisitions. With fading prospects of rental growth, the level of initial yield demanded by investors rises, causing capital values to drop even further. Thus, the property investment cycle has run its full course as the increasingly wary herd of investors turn to fresh pastures. This may involve switching into equities or gilts, or into different types of property unaffected by the downturn, or into property in other regions or abroad.

Inevitably, numerous developers are unable to secure a tenant or to sell the building either to an investor or an owner-occupier. The development sector enters a period of downturn. New development schemes, perhaps after years of painstaking site assembly, are curtailed, prolonging the blighting presence of buildings which have become derelict. Some half-built developments may even be temporarily abandoned, their completion awaiting the next upturn in the cycle after the over-supply of has been absorbed and some evidence appears of an up-turn in the economy and growth in user demand. Any new projects undertaken during the downturn tend to be either pre-let or pre-sold, or undertaken by owner-occupiers themselves. Few are speculative. Those which do occur tend to be concentrated in relatively risk-free prime locations.

2.7.6 Consequences

It is often said that the three most important factors impinging on the success of any property development are location, location and location. However, it is clear that getting the timing right is also important. To some extent, correct timing may be less crucial to an institution or large-scale commercial developer able to draw on considerable resources. But to a smaller-scale developer, bringing a project to the market at a time of oversupply can have ruinous consequences. For this reason, developers understandably become highly irritated by what they view as unnecessary delays in the planning process which can upset the whole timing and marketing of a scheme.

For some developers, the downturn can spell commercial ruin because their short-term debt still has to be serviced even in the absence of rental income. Thus,

> between 1987 and 1989, the total market capitalization of property companies on the New Zealand stock exchange declined by 78 per cent from NZ$5,782 to NZ$1,278 million. Of the top 20 property companies operating in 1987, 11 were in receivership, defunct or subject to takeover by 1989. (Moricz and Murphy, 1997, 177)

Similarly, the shares in half the property companies quoted on the London stock exchange lost over a quarter of their market value during the first half of 1990 alone (Fainstein, 1994).

Some developers fail spectacularly. New York-based developer Donald Trump, with debts estimated at over $1 billion, was obliged to pass control of his assets to his creditors, which involved over 60 syndicated banks (Fainstein, 1994). In the UK, Godfrey Bradman, an accountant who transformed Rosehaugh, originally a tea-shipping company, into one of the most innovative and successful commercial developers of the 1980s, saw company assets rise from St£18 million in 1984 to St£331 million in 1988. However, despite the declining market, assets had to be sold off in order to meet debt repayments. In 1991, he lost control of the company, which continued for a further year under a new management team until, in late 1992, Rosehaugh was put into receivership with liabilities of St£800 million and assets of only St£100 million (Adams, 1994). Similarly, in 1992, Speyhawk faced mounting financial problems with debts of St£300 million owing to 46 banks and went into receivership the following year.

Most remarkable of all was the fall of Canadian-based international developer Olympia and York, an interesting review of its downfall being provided by Fainstein (1994). The company, which once controlled some 2 million sq. m. (22 million sq. ft.) of office property in New York alone and had been responsible for such landmark developments as Battery Park City and was developing London's Canary Wharf docklands, faced a liquidity crisis. The withdrawal of long-term investors from the declining real-estate markets left it unable to sell its completed developments and unable to refinance its short-term development borrowings, secured against its portfolio of tenanted properties. It also found great difficulty in securing new occupiers in an over-supplied market characterized by declining demand after the 1987 stock-market crash which had occasioned widespread redundancies in the financial services sector. With real-estate borrowings of $12 billion syndicated between 90 and 100 banks, declining values in its

Figure 2.28 *Canary Wharf under construction, 1998. The public sector may promote and financially support redevelopment of industrial and dockland areas*

property portfolio and interest on Canary Wharf alone piling up at the rate of St£40,000 per hour, estimates of Olympia and York's asset shortfall ranged from $6–10 billion (Adams, 1994; Fainstein 1994).

Ball (1994) notes that periods of overbuilding, the development of the wrong types of structure or building at the wrong locations are inherent tendencies in property development. He further points out that 'buildings that turn out to be investment mistakes do eventually get completed and used for something, but not in the way, nor at the price, envisaged by the original investors' (Ball, 1994, 673). One example is a speculatively developed office block on Mountjoy Square in Dublin, in a decidedly marginal office location about 1 km. from the core. It was tenanted only after several years of vacancy and put to use as a further education college at a rent around one third the expected level. A neighbouring office development (Figure 2.29), which had remained half built for a decade, was eventually completed as apartments only after the introduction of tax incentives for property-related regeneration (MacLaran, 1993).

Figure 2.29 *Mountjoy Square, Dublin. Abandoned office development. The building was eventually completed as apartments after the introduction of fiscal incentives for property-led regeneration*

2.8 The impact of global capital flows

The global character of investment flows and their volatility, particularly during the increasingly deregulated financial environments of the 1980s and 1990s, can significantly exacerbate cycles, deepening the recessions and heightening the booms (see Dehesh and Pugh, 1999, 2000). First, an influx of credit increases competition between lenders, reduces the cost of finance and alters developers' profitability calculations in a positive manner, thereby encouraging new development. Second, overseas demand for investment property further stimulates development by competitively driving up prices for completed buildings. For these reasons, cheap credit and an influx of investment capital, became important catalysts of the UK property boom of the 1980s (see Nabarro and Key, 1992; Healey, 1994; Coakley, 1994). During 1988 and 1989, foreign investment in UK property amounted to St£3.1 billion, of which 44 per cent was of Japanese origin (Budd, 1992).

From his analysis of the impact of capital market operations on commercial real-estate development in the USA, Warf (1994, 325) has been led to conclude that:

> ... it is clear from the surge of investment-driven construction that swept the USA in the 1980s that finance capital is not some passive actor in the construction of landscapes, but an active participant with a logic of its own ... In this respect, commercial real estate markets reflect the endemic characteristics of capitalism in general: periodic overinvestment, oversupply and depressed earnings that define the inherently cyclical nature of capital accumulation.

However, while international deregulation can exacerbate the degree to which property markets overheat, the tendency towards overheating is an inherent feature of the industry and pre-dates widespread financial deregulation. Furthermore, it is not only the slow pace of development response to rising demand, nor the investment cycle that contributes to the over-supply of buildings. Intervention by the public sector, which lacks a market basis to its decision making, can also have a significant impact. For example, a combination of low real interest rates and the promotion of industrial property development by the Irish Industrial Development Agency during the late 1970s and early 1980s, resulted in a 10-fold increase in the completion of industrial floorspace in Dublin between 1977 and 1978 (MacLaran, 1993). Over the following four years, the enormous scale of development created such an oversupply of industrial space, estimated at over 500,000 sq. m. (5.3 million sq. ft.), that it thrust the industrial property development sector into a depression which lasted until 1989.

2.9 The impact of development cycles

The impact of the development cycle on the urban landscape has already been noted, particularly where dereliction is prolonged as developers await an upturn in user demand. However, there are major consequences for the property development sector itself, notably shifts in the balance of advantage between developers and the other key interests.

2.9.1 The developer–landowner relationship

Through the course of the development cycle, the balance of advantage is reflected in movements in the price of land and redevelopment sites. However, this relationship may be fairly loose. After an extended period of land price increases during a boom, landowners may be reluctant to revise downwards their price expectations during the slump. Conversely, during the early phases of a development boom, landowners may be unaware of the onset of accelerating development activity and sites may be acquired relatively cheaply.

Although the monopoly right over land confers considerable power to landowners, disposal of sites may be less straightforward than one might expect. Certain types of owner may have little capacity to adjust the timing at which their sites are brought to the market. For example, industrial operations or public institutions, such as hospitals, may decide to relocate based on grounds that have little to do with property considerations. The disposal of the original site and buildings, especially during a slump in the development cycle, can then prove problematic.

2.9.2 The developer–financier relationship

The impact of property cycles on the relationship between the developer and the short-term financier is essentially of an indirect nature insofar as user demand for buildings and the scale of demand for credit both reflect underlying trends in the economy. These will have an impact on both the cost of borrowings and the enthusiasm of the developer to initiate development. At times of credit shortage, financiers can win larger shares of development profits by increasing the cost of borrowings. The character of the relationship may also change as financiers effectively buy into development schemes in order to gain direct access to development profits by obtaining a right to receive a proportionate share.

The impact of boom and slump can be serious for the financial sector itself. In the USA, banks held over $400 billion of commercial real estate loans by 1990, the real-estate exposure of the Chemical bank and Manufacturers Hanover leading to their merger in 1991 (Fainstein, 1994). In the UK, the over-exposure of financial interests to the performance of the property sector, became apparent during the property crash of 1973–4. Over-eagerness to extend lending facilities to commercial developers at times of development boom had seen investment in the property sector quadruple between 1971 and 1973 at a time when the money supply had risen by 50 per cent. The ensuing crisis, saw prime office rentals in the City of London falling from St£220 per sq. m. (St£20 per sq. ft.) in 1973 to just St£120 (St£11) in 1976. This caused a number of secondary banks to fail and threatened to bring down the whole commercial banking sector. Ironically, to bail out the sinking ship of British finance capital and prevent the market from becoming flooded with the greatly devalued property assets of property companies put into liquidation, the Labour government devised a St£2,000–3,000 million 'Lifeboat Scheme' co-ordinated by the Bank of England. Between 1974 and 1978, some St£2,000 million worth of property assets were sold off to the financial institutions (see Boddy, 1981; Reid, 1982; Smyth, 1985).

Yet, by the 1980s, the UK property boom again saw massive increases in lending to the property sector, rising from St£2.2 billion in 1980 to over St£36 billion by 1992 (Pryke, 1994a; Coakley, 1994). The major difference was that overseas banks, particularly of

Japanese, North American and Swedish origins, were playing a far greater role and UK banks generally avoided the same depth of real-estate exposure. Nevertheless, Barclays and Lloyds banks were reputed to have lent $315 million and $100 million respectively to Olympia and York alone. However, these amounts were dwarfed by the exposure of the Honk Kong and Shanghai Banking Corporation at $750 million, the Canadian Imperial Bank of Commerce at $713 million and several Japanese banks whose loans to O&Y totalled $780 million.

But, as Ball (1994, 689) observed, it is 'unlikely that another future property boom will be held back by a lack of credit; not simply because lenders forget the lessons of the past' but because the factors which arise in the new situation, such as new sources of credit and innovation in development financing, make those lessons irrelevant.

2.9.3 Relationship between the developer and construction interests

Both extremes in the property development cycle bring problems for the contractor. While it is easier for developers to refrain from initiating new developments, it is vital for construction companies to maintain a relatively unvarying flow of work. At times of slump, contractors face greater competition for available contracts and their profit margins may be severely squeezed by developers, who may even be able to hold builders to fixed-price contracts. Desperation to win contracts may induce contractors to submit unrealistically low tenders. If successful, these frequently result in lengthy disputes with the client over inflation-related increases and cost adjustments for any alterations to building specifications. Furthermore, as development activity takes off and demand for building services and materials rises, contractors may then be faced with rising prices of materials and increasing labour costs as skills shortages arise. Yet the contractor may be tied into a contract which may run for two years or more.

A major consequence of the development cycle is discontinuity of employment, which is felt throughout the sector from building labourers and skilled trades, to professional consultants, including surveyors, structural engineers and architects. During the mid-1980s in Ireland, when downturns in the development cycle were experienced simultaneously in housing, office and industrial property development, rates of unemployment in the construction sector rose above 50 per cent and many sought employment abroad. In contrast, during the boom of the late 1990s, when Irish economic growth topped 7 per cent per annum for several years in succession, demand for building services grew rapidly. Skills shortages soon emerged, wage rates rose rapidly and building workers were recruited from abroad.

Such instability is a major contributor to the problematic relationships between management and labour which seem endemic in the construction sector. It is also an underlying explanation for the proliferation of sub-contracting within the construction sector (Smyth, 1985). The use of sub-contractors, ranging from specialist services to unskilled 'lump labour' gangs engaged for the duration of the contract, minimizes direct employment by the main contractor and the scale of redundancy payments required when the downturn inevitably arrives. Similarly, plant is hired when and where needed in order to reduce the scale of fixed capital tied up in the business. Most recently, 'the most mature expression of minimizing overheads, constant and variable capital, is the emergence of management contracting. The main contractor will not necessarily undertake any of the site work … the entire contract being subcontracted to other agents' (Smyth, 1985, 77).

It is partly because of the precariousness of their situation that construction companies have sometimes sought to extend their sphere of activity into the development field to permit a more guaranteed flow of work and greater access to shares of the development profits. It can be achieved either by entering into development partnerships with commercial property developers, effectively buying into particular schemes, or as a longer-term strategy involving the establishment of separate development companies. However, the capacity of such initiatives to increase the total volume of work at a time of low user demand is severely limited, although by winning greater shares of development profit during boom periods, the construction sector could be better placed to weather the ensuing slump.

An alternative strategy to cope with downturns in development activity is to switch the focus of activity geographically to benefit from the different timing of development cycles regionally or internationally. In so doing, the construction interests have followed the example of the developers themselves. Generally, however, this proves more difficult for the industrial capital of the construction sector.

2.9.4 The developer–investor relationship

Because the prospective returns from property vary through the development cycle, institutional investors adjust their degree of involvement in property accordingly. But only a proportion of clients' funds will be available for investment in property as clients may themselves determine the nature of investment. They may opt for fixed interest risk-free government stock, for equity shares in companies or purchase into a property unit trust or bond. The quantity of funds being directed towards specific types of investment is therefore likely to reflect depositors' appraisals of short-term performance of different investments. The promises of the institution will also constrain the quantity of funds available for property investment, guaranteed income bonds requiring investment in fixed interest government stock. From those funds available for discretionary investment, only a proportion will be directed into property. Ideally, property acquisitions should be made when markets are weak but about to rise, benefiting from low acquisition costs but avoiding a lengthy period without rental growth. However, the low points of any market tend to be easy to identify only with hindsight!

However, the fact that large institutions receive a substantial proportion of their investment funds in the form of life insurance premiums or pension contributions, provides a continuous flow of money deposited for the long term. This reduces the pressure for immediate investment performance. Matching long-term investment with relatively predictable long-term liabilities, they are somewhat removed from immediate market pressures and are better able to withstand short-run market vicissitudes. This confers on institutions considerable 'money muscle' and places them in a strong market position when dealing with debt-laden commercial developers working under the imperative of making interest payments on borrowings from the short-term money markets while trying simultaneously to provide a dividend for shareholders. It may mean that acquisitions from cash-starved developers can be made cheaply, especially at times of downturn in the development cycle. Thus, the life insurance company Scottish Widows managed to acquire the Fleet Walk Shopping Centre in Torbay, England, once valued at St£40 million for just St£27 million in 1991 (Adams, 1994). Indeed, as a result of the office property crash in the UK, in the mid-1970s, not only were the institutions able to obtain large quantities of space cheaply, but they also absorbed many of the commercial development companies themselves (Franklin, 1976).

This 'money muscle' also favours the institutions when it comes to development. While correct timing of development is important for all developers, it is especially crucial to commercial developers for whom the failure to find tenants for a completed scheme may result in liquidation. Though still costly to an institutional developer, over a 60-year investment life of a building, the short-term vagaries of the accommodation market become less significant.

Yet, although it is possible to identify periods in which the institutions increase their direct development activity, at each new development cycle there seems to appear a new generation of commercial developers, often more adventurous and availing of innovative financial arrangements (Key, Espinet and Wright, 1990). Ease of entry into the development sector encourages such newcomers, the entrepreneurial developer often playing an important role as 'market openers' (Healey, 1998). Through their lower risk aversion, they are more likely to undertake schemes in riskier locations, as Pryke (1994a) showed for areas marginal to London's established financial district during the 1980s. Others may avail of changing organizational contexts, particularly locally-based developers who emerged to take advantage of new public–private sector arrangements and special incentive regimes for property-based urban regeneration, such as in Enterprise Zones (see Key, Espinet and Wright, 1990; Healey, 1994).

Inevitably, many trader-developers will fail while some prosper in a Darwinian process of survival of the fittest, comprising those who have greater ability or simply more luck (see Ball, 1994). By demonstrating flair, financial innovation, accurate market perception, a speed of response and an agility of negotiation, some prove able to outmanoeuvre the more ponderous institutional developers by gaining access to sites and operating without having to relinquish so much of the development profit to the institutional long-term investor giants. But even for successful entrepreneurial developers, O'Donnell (1989, 34–5) has suggested that there may also exist a natural lifespan:

> ... most development organizations seldom last more than 10 years. The pattern is familiar. Developers work very hard and enjoy fair success in their early years. They become accepted by the financial community, and their success increases, especially if they have caught an up-cycle. Then, strangely, their fortunes shift: either, believing they have the Midas touch, they risk all on several poorly planned transactions and lose, or they decide to conserve assets in the face of the personal liability risk. The newly conservative developer loses his or her more aggressive senior employees or partners and ends up as a manager of already developed assets.

The foregoing extended discussion has reviewed the character, motivations, operations and interactions of the major private-sector roles associated with property development. Attention must now turn to a consideration of how and why the state, through its urban planning procedures, attempts to intervene in this process of property development.

2.10 Notes

1. Objects do not possess an intrinsic characteristic of belonging; they have no inherent 'your-ness' or 'my-ness'. 'Property right' describes the social relationship between legally-defined individuals and things or, in slave-owning societies, between one individual and another (the object of ownership). 'Capital' is ultimately based on this concept of property right. Rather than its being an identifiable 'thing', capital is more appropriately understood as a relationship based on the power to exclude others from the use of the productive means of society, founded upon a privatized claim to control them.

   In developing his materialist critique of legal forms, Evgeny Pashukanis (1924) [1978] noted that the definition of humans as legal personalities and bearers of 'rights', rather than possessors of customary privileges, coincides with and is necessary for social production to be pursued as a process of general commodity production. Thus, 'only in bourgeois capitalist society, where the proletarian figures as a subject disposing of his labour power as a commodity, is the economic relation of exploitation mediated legally, in the form of a contract' (Pashukanis, 1978 edition, 45). It is therefore the commodity form which produces the legal form and 'the legal relation between subjects is simply the reverse side of the relation between the products of labour which have become commodities' (Pashukanis, 1978, 85). This is further elucidated in Arthur's (1978, 14) introduction to the 1978 edition:

It is only in the conditions of commodity production that the abstract legal form is necessary – it is only there that the capacity to have a right in general is distinguished from specific claims and privileges. It is only the constant transfer of property rights in the market that creates the idea of an immobile bearer of these rights. Indeed, the abstract capacity of everyone to be a bearer of property rights makes it difficult for bourgeois thought to see anything else than subjects of rights: legal fetishism complements commodity fetishism.

3
Planning the city

Andrew MacLaran and Pauline McGuirk

3.1 Introduction

The previous chapter identified the fundamental private-sector forces responsible for generating the changing form of urban landscapes. It examined the structural relationships within the property development sector, together with the motivations, activities and market-based constraints on the major interests. This chapter reviews the context within which those forces operate. It examines urban planning, which is here defined as comprising physical development planning involving the strategic regulation of development as a means of controlling the production and use of urban space. It therefore concerns the way in which planning affects the development process, land and property rights, real-estate interests and markets (Healey, 1997). It examines the ways planning intervenes, either through passive controls or, more actively, through incentives and inducements, to bring about specified outcomes which authorities deem desirable, rather than those likely to result from unfettered decision-making by individuals exercising their property and development rights in an entirely self-interested manner. Those outcomes deemed desirable by planners are often constructed around some notion of 'orderly development' in line with the general 'public interest', conforming with principles aimed at producing a 'functionally efficient' city or to some conception of 'social justice' in the allocation of resources. However, as we shall see, whether these outcomes can actually be identified is highly contentious.

The chapter sets out the ideological foundations of planning practice, its instruments and mechanisms of intervention and, in order to shed further light on the various theoretical interpretations of urban planning's structural role, it also reviews the conflicting theoretical perspectives on the role of the state in capitalism.

Finally, it discusses the relationship between urban planning and private-sector property developers, for which the creation of the built environment is a source of capital accumulation. This leads into a brief discussion of recent changes in urban planning practice towards greater degrees of pro-active operation, or 'entrepreneurialism'.

3.2 Urban planning as a state activity

Urban planning is a form of collective action undertaken by state authorities. It represents an attempt by the political state to influence, shape and control the results generated by the property development forces reviewed in Chapter 2. Traditionally, it has sought to influence the strategic planning and development of land and transportation infrastructure, the functions to which land and buildings are put, together with the form of development including its scale, architectural appearance and construction quality. Urban planning may also seek to address undesired results of the way urban landscapes developed in the past and are now considered inappropriate. This might involve dealing with 'incompatible' land uses, such as polluting industrial activities located near residential areas.

Thus, in the process of urban development, the urban planning system establishes the framework within which property development takes place. However, the authority of planning and the degree of power with which it is endowed vary from one political state to another according to broader economic, socio-cultural, constitutional and legal, ideological, political and administrative contexts. These create considerable differences in the character of urban planning from one jurisdiction to another.

Formal planning controls may be quite weak or even virtually absent, as was long the case in the past and continued to be common until very recently in many Texan cities. At other times or in other places, regulation may be tighter. Indeed, as will be shown later in the chapter, growing flexibility and negotiative approaches have been adopted increasingly by planning authorities under pressure to attract and facilitate economic growth in their localities. Thus, planning has tended to become part of a generalized process of strategic urban development, providing a guiding role in response to the demands of the urban 'boosterist' lobby. The case studies of Birmingham and Sioux Falls in particular show how urban planning has assisted in a municipal development strategy involving the re-creation of both the image and reality of the city, helping to create an urban landscape which is iconographic of economic success, growth and change.

The range and detail of goal setting in urban planning may also be influenced by the degree to which planners' political masters believe private property rights should be constrained and collective action promoted. However, as the case studies of both Minneapolis and Sioux Falls show, urban planning can be highly interventionist even in predominantly *laisser-faire* economies, their predominant planning ethos being overtly facilitative of the development sector. Similarly, the instruments available to urban planners also vary considerably, their character and potency again being determined in large measure by the way in which the political establishment regards state intervention and its underlying purpose.

Furthermore, there are numerous different legal bases to planning, which determine the degree to which plans have statutory force or whether they are merely indicative and open to negotiation between developers and planners. These have been excellently reviewed for European countries by Newman and Thornley (1996). Thus, in contrast to the flexibility of British urban planning, where 'appropriate' development can be a subject of considerable negotiation between developers and urban planners, the Italian system is based on legal prescription of what is permitted, allowing little flexibility and only marginal adjustments to prescribed outcomes. Recently, in some jurisdictions, there has been a growing trend towards increasing flexibility over the control of land uses by introducing to the zoning system extensive categories of 'exempted' or 'complying' development which are automatically approved (McInerney, 1998). This theme is explored further in the Sydney and Auckland case-studies.

3.3 Planning practice

The instruments of planning commonly include permissory elements embodied in guidelines, controls, regulations and codes. These include tools such as the zoning of permitted land uses, development controls determining permissible plot ratios and building heights, and the design, construction characteristics and architectural appearance of schemes. The protection of existing urban landscapes is also effected through the conservation listing of individual buildings or whole districts. Taken together, such controls represent an expression of policies aimed at ensuring that only development of an approved type can take place. However, these are largely reactive measures, comprising a passive planning framework whose impact and effectiveness is dependent on applications for development permission being forthcoming. At times of slump in property development when few applications arise, planners' powers become constrained.

In an attempt to intervene more actively in property markets, the passive elements might be augmented by a variety of incentives. These may involve stimulating property development in general by providing incentives to increase user-demand for buildings. Alternatively, they may attempt to alter the configuration of the 'profitability surface' for property development by allowing greater development densities in specified 'riskier' locations where planners seek to promote development. For example, allowing greater plot ratios helps the redress the problem of lower rental values in certain areas.

More direct intervention may involve the provision of fiscal incentives or financial inducements for development, commonly being restricted to qualifying districts ('designated areas' or 'enterprise zones') or to individual redevelopment sites. Such inducements have become an increasingly common feature of more 'active' modes of urban planning, associated with deepening levels of entrepreneurialism in urban governance (McGuirk and MacLaran, 2001).

As noted in the Introduction, the state can also intervene in an even more direct fashion by taking on the development role. This applies especially to those urban elements that are beyond the logic of the private sector to provide, such as social housing and the provision of infrastructure such as roads, sewerage and amenity space. However, as this text concerns the relationship between planning and private-sector development, these public-sector operations are not further discussed.

3.4 Shaping the urban landscape: the genesis of urban planning

Political states have long regarded control over urban development as an important aspect of their role. From the time of the first cities, there is evidence of the organization of land-uses and the co-ordinated provision of basic infrastructure, from regularly laid-out streets to piped water and drainage systems. These reflect developments in hierarchical social organization and control associated with the rise of class-based societies and the emergence of political states.

The division of labour and the hierarchical organization of functions became reflected in the separation of functional areas; the palace precinct, the religious buildings and the market place being identifiable components in the early organization of urban space (Mumford, 1961). With the physical separation of the monumental buildings of the élite from the densely developed housing of lesser mortals, urban space additionally took on an ideological role. Thus, the urban landscape came to embody, symbolize and reinforce the differences in social power, both reflecting and helping to mould social relations.

This relationship between urban and social form became ever more complex with subsequent historical development. However, as Houghton-Evans (1978, 190) has suggested, 'we need not look far ... for evidence of the way propertied ruling classes throughout history have shaped the city to fulfil their political and ideological objectives'.

Figure 3.1 *The 60-storey Woolworth building on Broadway, Manhattan, was described on its completion in 1913 as a 'cathedral of commerce'*

The modern capitalist city is no exception, the ideological foundations of capitalism becoming intimately inscribed in the structure and forms of the city (Domosh, 1992). In the same way that the medieval cathedrals of the Roman Catholic Church affirmed its authority, so skyscrapers, the cathedrals of modern commerce, proclaim the power and pretence to permanence of the corporations which occupy them. Similarly, the degraded urban environments of the urban poor reflect and perpetuate their disadvantage, marginalization and feelings of powerlessness. (Discussions of the connection between economic power, social control and urban space in Los Angeles and New York are respectively provided in Davis (1990) and Smith (1996).)

Urban space itself is partitioned and subject to 'rights' of ownership and control. Our urban lives are played out on a geographical stage of such property rights: from the individualized private space of the residence, through state-owned 'public' spaces connecting us to other private spaces of work and leisure. The structure of our operations, our forms of behaviour and the conditions under which we use those spaces demand our complicity and 'naturalize' the notion that this landscape of property rights and, indeed, the very notion of 'property rights' themselves are legitimate and inevitable.

Just as state intervention in urban development is longstanding, so utopian theorizing about the character of the 'ideal city' also possesses a lengthy pedigree. It embraces discussions in classical antiquity about the most appropriate population size and aesthetic considerations of architectural form (e.g. Plato and Vitruvius) and includes artistically designed street plans devised for laying out the ideal Renaissance city (see Hiorns, 1956; Burke, 1971; Morris, 1972). However, the history of urban planning is not a simple linear story of the gradual 'discovery' and application of a pre-ordained set of planning principles. Cities inevitably mirror the societies that engender them. Planning evolves as a reaction to and out of a concern over the way conflicts and problems inherent to the structure and operation of the social system tend to become reflected in the built environment As Scott and Roweis (1977, 1100–1) have cogently observed:

> We cannot justifiably make the *a priori* assumption that planning is simply an independent and autonomous sphere of social activity. That is to say, we cannot assume that urban planning emerges, acquires its observable qualities and evolves, according to forces that reside *solely within itself* … it is not an independent and autonomous planning *theory* that produces the various facts of actual planning: it is rather the realities of contemporary urbanization that give rise to planning as a necessary social activity. Subsequently, a constellation of theoretical ideas … is built up around this necessary social activity.

3.5 Planning ideology

Gradually, the modern state became intimately involved in the regulation of private-sector development. However, the specific emergence of modern town planning out of a coalition of interests which included sanitary reformers, the philanthropic housing movement, the garden city movement, countryside preservationists, urban architectural conservationists and, in the USA, the City Beautiful movement, conferred upon urban planning a particular and modernist ideological inheritance. The ideological foundations of these interests became embedded in the practice of urban planning through the prescription of standards and, with the adoption of codes and technical formulae, in the promotion of regularity and order.

Knox (1982) identified five important components of traditional urban planning practice resulting from these origins, which continue to infuse many aspects of the practice of urban planning either in strategic goal-setting or when dealing with development applications:

1. Environmental determinism: the idea that it is possible to improve the physical, moral and social welfare of people by up-grading their physical environment led to planners' concern for prescribing standards and formulae for achieving environmental and building quality.

2. Aesthetics: planners' concern for the creation of visual order, harmony, scale and townscape can be traced to the civic design or architectural influence on modern planning and can be seen in controls over building height, design, massing and alignment which affect the townscape

3. Spatial determinism: the concern for separating urban functions through the practice of zoning, in order to eliminate incompatible land uses, derives from the spatially deterministic assumption that spatial order brings social and economic benefits. Similar assumptions underlie the concept of the 'neighbourhood unit', encapsulated within the concept of 'new urbanism' which promotes the idea that careful layout of residential areas leads to social interaction, the development of 'community' and the stimulation of feelings of security and stability (Talen, 1999).

4. Systems approaches: in attempting to introduce techniques such as operational research, systems analysis and cybernetics in the monitoring and modelling of the urban environment, planning appears as a technical process. However, while the mathematics of modelling may be entirely correct, the assumptions upon which such models are based, the forms and sources of the data which are drawn upon and the criteria for evaluation of alternative options, depend on goals which are inherently political (see below).

5. Futurism: planning has long been imbued with a penchant for futuristic idealism. But the vision of the 'ideal' and the values inscribed within it tend to be based on a particular culture from which the planning profession draws its functionaries. It is a vision that assumes middle-class lifestyles, values and behaviour are aspired to and attainable by all and that their achievement would benefit everyone. However, such values have tended to be based on the class, race and gender visions of the dominant culture and have reflected a Euro-centric and masculinist worldview, sense of identity and aesthetics. Moreover, in this critical light, the low-density suburban form as an idyll to which all are assumed to aspire engenders an ideology of possessive individualism which stimulates households' demand for commodities of every type (see Harvey, 1985).

Contemporary critiques of planning ideology point to these elements as reflecting planning's modernist heritage (see Milroy-Moore, 1996; Healey, 1997; Sandercock, 1998). Born of the Enlightenment of the late eighteenth century, modernism is underpinned by a positivist epistemology (or theory of knowledge) which privileges scientific technical knowledge and empirical data as the only valid and convincing form of knowledge. It therefore involves a strict separation of verifiable fact, assumed to be deduced through reason, from more transient values, assumed to be based on 'mere' emotion, aesthetics or morality. The techno-scientific reasoning associated with this form of knowledge has produced the view that reality can be controlled, ordered and perfected once its underlying laws and objective logic can be uncovered.

Thus, modernity is based on a faith in the ability, through scientific and technical reasoning, to achieve progress and betterment, which is assumed to be universally advantageous. This view is possible because modernism assumes humans to operate as individual, self-interested and autonomous beings who are fundamentally similar; encountering and experiencing reality in essentially the same manner, developing a common sense of being and operating according to functional rationality and predictable sets of social laws.

The aim of progress embedded in the rise of modernism is closely intertwined with the development of capitalism and of the capitalist state (Beauregard, 1996). Indeed, the state has been the source of the systematic organization, ordering and management of social, political and economic life considered necessary to achieve a rational and efficient allocation of resources. In other words, modernism has positioned the state as a central rational force in achieving progress and Modernist principles, particularly its techno-scientific rationality, have been embedded in many of the decision-making institutions and bureaucracies of the state. Urban planning is no exception. The profession has been built on the premise that well-ordered and efficient cities could be created through the application of uncovered laws of spatial behaviour and organization (Johnston et al., 1994).

Sandercock (1998) has identified five pillars of modernity that have been foundational in the development of urban planning practice.

1 Rationality
Scientific rationality's basis and its emphasis on 'verifiable facts' has led it to be viewed as a superior means of public decision-making because it could remove any political prejudice or ideological taint. The administrative systems of planning and its technical tools and instruments (e.g. functional categorization into activity zones) have been developed according to scientific rationalism (Healey, 1997). Thus it is assumed that social problems or inefficiencies in land uses or in the allocation of resources could be solved by the application of logic.

2 Comprehensiveness
The evaluative tools and mechanisms of urban planning are founded on a particular view of cities as integrated systems of economic and social relations that are amenable to

coherent management based around scientifically constructed knowledge and technologies. Thus it is believed that the city can be comprehensively physically planned according to logically-derived laws of urban development to produce an orderly and spatially integrated city that fulfils the functional necessities of urban life. This has been the basis of the so-called 'sound planning principles' from which many planning decisions and policies have been derived.

3 Scientific objectivity

By virtue of their expert knowledge and scientific approach, planners have claimed a critical distance from political views and values, and from the conflicting interests of capital, labour and the state itself (Beauregard, 1996).

4 State action as a route to progress

Modernism has tended to envisage the state as a progressive institution with the ability to organize and implement reform. The state tends to be viewed as a political field, separate from and without vested interest in the economic sphere. It is therefore regarded as 'a benign agency of implementation of masterplans' (Sandercock, 1998, 23). State planning thus becomes a vehicle for progress.

5 The public interest

The notion of 'the public interest' denies that there might exist between social groups a fundamental divergence in social values. Rather, it assumes that despite superficial differences, social groups are united by shared interests, needs and values. Differences of class, race and gender are ultimately not considered relevant. Planners are assumed to be capable of identifying those shared interests as the 'public interest' and of operating as apolitical mediators in conflicts of interests in the name of 'the common good'. Planning therefore functions to create physical environments which embody and accommodate the assumed public interest.

While these assumptions have long been questioned in critical reviews of planning theory and in discussions relating to social theory more generally, many remain central to planning practice nonetheless.

3.6 Conceptions of the state and the role of urban planning

As urban planning is a function of state activity, our understanding of its existence and our explanations for its operations depend upon how we view the role of the state in capitalism. Kirk (1980) has provided an excellent review of four of the major interpretations of the state's role: pluralist, managerialist, reformist and Marxist political-economy perspectives. Although these may initially seem irreconcilable, each possesses a certain degree of validity and offers a different depth of insight into the relationship between the political process and social structures. Taken together, they provide a nested hierarchy of explanatory power.

Pluralism stresses the ability of everyone to participate in and influence collective decision-making through the political democratic process. However, managerialism recognizes that bureaucracies may follow their own agendas, become obstinately attached to favoured policies and prove difficult to influence. Marxist interpretations of the state attempt the most penetrating and all-encompassing analyses. They stress the state's ultimate structural role of guaranteeing the privatized ownership of the means of production and the continued social and economic domination by capital. This it accomplishes through maintaining the social relations necessary for the reproduction of the capitalist system. Reformist perspectives point to the possibility nonetheless of the state's undertaking marginal reforms to effect redistributions within the confines of that system.

These four conceptions are now considered in some detail. This section also provides a review of postmodernist critiques of Marxian political-economy approaches, the alternative conceptions of planning which they promote, together with some cautionary observations regarding their insights.

3.6.1 Pluralism

Pluralism maintains that society comprises a range of different legitimate and conflicting interest groups with a diversity of values (Dahl and Lindblom, 1953). These groups vie with one another for influence over government policy. However, no single group is able to dominate as power is distributed across social groups in a diffused manner. The prime function of the state is to achieve consensus, work to effect social stability and maintain a rough balance of power between interest groups by accommodating its actions to the variety of competing interests. Thus, state decision-making and policy directions are regarded as determined by and effected in response to the relative strengths and balance of pressure between competing interests. The state acts as an arbiter between such groups, supervising and regulating them to ensure that none gains mastery and dominates state policy, thereby undermining the illusion of political democracy and threatening the perceived neutrality and legitimacy of the state's operations.

In the arena of urban development, many interests vie to be heard by those with planning power. These typically include conservation groups and environmentalists, community and residents' associations, transportation lobbies and a well-organized 'property lobby' encompassing property owners, developers, construction firms and numerous professional institutions including surveyors, architects and engineers.

While the pluralist thesis acknowledges that conflicting values and interests exist, it maintains that these can be accommodated within the contemporary societal structure. It emphasizes the ability of people to organize around issues which concern them, that different interests each have adequate power and access to decision makers to enable their opinions to be voiced and assumes that bureaucrats are receptive to such appeals. Thus, pluralism is an inherently conservative perspective, viewing state policy as a 'fair' and 'balanced' outcome of competition between different groups. This verges on a market theory of political activity (Broadbent, 1977).

Under this perspective, which still holds considerable sway in planning circles, planners are viewed as 'referees' and 'power brokers', guardians of 'the public interest', seeking to resolve conflict between competing interests to serve the 'common good'. Indeed, public consultation and participation in the practice of planning has been viewed in planning theory as a means of improving planning outcomes through practising pluralism. Pluralists might therefore contend that more inclusive pluralist policy outcomes could be developed by providing forums in which the conflicting views of various interest groups could be worked through and resolved through planners' adopting 'proper planning' solutions.

However, pluralism has been criticized for failing to recognize the enormous inequalities in resources between individuals, communities and classes in terms of their economic power and the organizational skills upon which they can draw. While some are well funded, well connected, well informed and equipped to act effectively in their own interests, others are impoverished, fragmented and marginalized, lack appropriate information and contacts or the social and political resources to enable adequate mobilization. Thus, in contemporary capitalist systems characterized by an absence of 'economic democracy', marked by considerable inequality in economic power over material resources and the means of living, the imagined 'political democracy' of pluralism may justifiably be regarded as a fiction.

Pluralism has equally been attacked for assuming that a lack of political activity by citizens reflects satisfaction rather than ignorance, frustration, apathy or disillusionment over political action, which might have the effect of disempowering people. Moreover, it fails to take account of what is often termed the 'second face of power'; the ability to limit the scope of the political process to the consideration of issues which are generally innocuous and which pose no real threat to the structural bases of economic power.

Critics of contemporary public participation in planning have argued that, in the absence of recognition of inequalities in access to economic, cultural and political resources, pluralist participation serves merely to impose on communities the technical and scientific rationality characterizing the practice of planning (Gleeson and Low, 2000). Public participation based on a pluralist conception has also been criticized as a means of establishing a notion of a 'public interest' or 'common good' rising above the diversity of values and interests of a multiply-fractured society. The notion of planning as the pursuit of the public interest has promoted a professional belief that planning can be an objective, apolitical activity serving common values and providing mutually beneficial outcomes for all.

Arnstein (1969) presented a landmark critique of public participation as a means whereby groups' interests could be taken into account by decision-makers, such as planners. Her work undermined the idea that planning operated as an apolitical, pluralist arbiter of competing claims, or indeed that pluralist participation was possible in the presence of vast inequalities in the resources and strategic abilities of interest groups. She also recognized a vast gulf between community resources and those of powerful planning authorities.

In giving due recognition to these inequalities, Arnstein equated genuine participation with the redistribution of power between interest groups through negotiation. She presented a hierarchy of gradations of citizen participation, ranging from the non-participation of 'manipulation' and 'therapy', to the tokenism of 'informing', 'consulting' and 'placating', to the empowering participation of 'partnership', 'delegated power' and 'citizen control'. She observed that participation does not provide a pluralist road towards finding a 'proper planning solution' serving the common good. Rather, it is a political power game that can involve the manipulation or co-option of group demands towards courses of action preferred by planners. Furthermore, her work implied that what could be achieved through participation could vary tremendously depending upon the resources held by the interests groups involved. Rather than comprising a means of effective democratic pluralism, public participation in planning could be wholly 'meaningless if only the more powerful sections of the community are involved', as Watson and Gibson (1995, 257) have suggested.

3.6.2 Managerialism

Managerialism cites the complex and impenetrable structure of the state machinery which renders it impervious to influence from its citizens (see Davies, 1972). People may be ill-informed, apathetic towards, or ignorant of their legal entitlement to participate. When faced with complex bureaucracies, they may be unaware of to whom they should most appropriately address their concerns or how most effectively to make their opinions heard. The professionalization of decision-making and its reliance on expert knowledge systems, together with the often inaccessible jargon, technical language and norms of communication adopted by bureaucracies, create barriers to discourage the uneducated from effective participation. As one planner interviewed by McGuirk (1991, 266) cynically commented about the procedures for public participation in drawing up development plans, 'we are expert enough to make it complex enough for the average person not to understand'.

Furthermore, in public bureaucracies operating according to expert knowledge systems and techno-scientific forms of reasoning, the evaluation of competing claims is likely to be based on a particular definition of 'rationality' which does not recognize as valid those arguments based on emotional or moral reasoning and values (McGuirk, 2001). Direct contact with officials and managers may therefore fail to deliver satisfactory results.

Meanwhile, managerialism maintains that elected council members, who provide a less direct route for citizen participation in planning, become dominated by bureaucratic rules and reliant on the expertise of the professional bureaucracy. This renders them insensitive to the needs of their own constituents. It further maintains that careerism and empire building, in which managers seek to promote their own interests rather than pursuing apolitical and technical management functions in their respective policy realms, are typical of bureaucracies. This bureaucracy includes planners, who operate in a managerial capacity. For example, despite policy changes to favour public transport systems, highway engineers have often proved reluctant to abandon long-cherished plans for urban motorways or road widening projects to accommodate more private cars. Such policy shifts threaten embedded bureaucracies, with their established career structures and administrative power bases and change may be long resisted, sometimes in ways amounting to subterfuge or outright subversion of public policy.

Under the managerialist interpretation of the state, the power of public bureaucracy is stressed. Insofar as they possess considerable power over the allocation of real environmental resources, planners are viewed as social 'gatekeepers'. For some urban localities, this may mean the provision or preservation of advantageous elements, such as a public park or the protection of views. For other communities, it may mean bearing the costs of proximity to waste tips or urban motorways.

However, the implicit assumption of much of the literature which adopts a managerialist perspective on the role of the state and urban planning is that 'bad' planning could be avoided if only bureaucracies were made more politically accountable and receptive to public participation in decision-making. Although local state bureaucracies can indeed be impenetrable, critics of the managerialist perspective have therefore pointed to its theoretical weaknesses, arguing that it fails in the same manner as pluralism to recognize the fundamental limitations to state activity, both centrally and locally. Like pluralism, managerialsm conceptualizes the political realm in isolation from its contextual economic base. These therefore stand in sharp contrast with the perspective provided by Marxist political economy which emphasizes the limits to state intervention, wherein the state's very aims, together with the range and depth of its operations are structurally constrained.

3.6.3 Reformism

The reformist interpretation of the state stresses its role as a vehicle for social reform. It recognizes the inevitability of inequality in capitalist society and that the state, whilst helping to maintain the viability of the economic system, pursues courses of 'corrective' action intended to ameliorate its most undesirable outcomes. This it effects through implementing welfare policies aimed at the redistribution of real income between groups, places or time periods. This perspective recognizes the transformative potential of the state, viewing it as having an inherent tendency to foster reform (Beauregard, 1996). It regards the state not as a monolith working in the service of capital but as a terrain of political struggle, separated from the economic sphere, possessing the potential for generating progressive policy outcomes to address social need.

Planning, as a function of the state, can thus be viewed in a reformist light, revealing a strong faith in its transformational powers. Indeed, this is an interpretation common among planners themselves, many of whose interests include more than purely physical planning but extend to a concern for urban social equity. Many planners have been attracted to the profession by its social democratic vision of 'improvement' and through a belief that planning might produce a more 'equitable' city by operating as a 'progressive' means of redistributing urban resources away from urban élites towards the disadvantaged. The traditions of advocacy planning (Davidoff, 1965) and equity planning (Krumholz, 1994), in which planners became engaged particularly with disadvantaged communities to assist them in formulating and achieving planning outcomes in their favour thus highlighted distributional questions in planning practice. Metropolitan strategic planning has also traditionally been concerned with attempting to contain and redress the inefficient and socially inequitable distributions of resources, goods and services which can emerge from the market (Gleeson, 1998).

Though weak in terms of its theoretical derivation and its conceptualization of the limits of state activity, it is evident that the state does attempt to effect marginal redistributions between competing interests. Thus, in return for favourable development permissions, urban planning may secure marginal redistributions, 'community benefit' or 'planning gain' which divert some share of the benefits accruing from development towards lower income groups. For example, in 1984, Boston introduced 'linkage' schemes between real-estate development and community benefits. One scheme obliged commercial office developers to pay into a housing trust fund a sum of $5 for every square foot (c. $50 per sq. m.) of development above 100,000 sq. ft. (9,290 sq. m.). By 1988, this had raised $45 million and assisted the development of 1,700 dwellings, 80 per cent of which comprised 'affordable' housing units (Coyle, 1988).

Similarly, under the New South Wales state-government-sponsored redevelopment of the peninsula of Ultimo-Pyrmont, immediately west of Sydney's CBD, a development levy was imposed on residential development sites. This obliged developers to contribute 0.8 per cent of their floorspace or to make a contribution of A$16 per sq. m. (A$1.48 per sq. ft.) of development (NSW Department of Planning, 1994). The funds were used to develop affordable housing to ensure that some low income housing was maintained on the peninsula, which would otherwise have been driven out by the high market value of waterfront residential land located so close to the CBD.

Nevertheless, it is clear that the degree to which urban planning is able to effect such redistributions in capitalist cities is constrained by the imperative to maintain private-sector development profitability. If this were jeopardized by the scale of required planning gain, it would be highly unlikely that development could proceed.

Reformist conceptions of the state have re-emerged recently, embodied in the implicit stance of postmodernist critiques of Marxian political economy (see below).

3.6.4 Marxist political economy

Political economy has a number of variants but all are founded on the notion that the economic and political spheres are inextricably linked. The Marxist political economy perspective pays particular attention to the role of the state and provides a penetrating analysis of its relationship with its societal base. It does not render the three previous explanations entirely redundant or irrelevant. Each has some observable validity. Rather, within the 'nested hierarchy' of explanations referred to earlier, it provides the most profound conceptualization of the state's origins, its evolution and contemporary functions.

Marxism regards the state as a social institution which arose historically in order to assuage the potentially inflammatory conflicts inherent in the emergence of class-based society. Its specific form and social role are therefore rooted in the mode of production. In capitalism, the inherent conflict between the interests of capital and labour over the division of social product threatens the basic social (including property) relations of the capitalist mode of production. Because capitalist society is unable spontaneously to ensure the successful reproduction of conditions for its own existence, the state acts to assuage the conflict between capital and labour, to mediate class conflict, to legitimate capitalist society and its property relations and, ultimately, to guarantee the relations of production. Thus, there develops an intimacy of relationship between the economic foundations of society and the political and ideological superstructure.

Marxism therefore conceives of the capitalist state as a creature of the requirements of capitalism, undertaking those operations which are imperative to the continuous operation of the system but which lie beyond the logic or profitability criteria of any individual elements of capital to perform. In other words, the operations of the state ensure a sufficient degree of economic and political harmonization and ideological control to enable capital accumulation (the driving force of capitalism) to be sustained. Even though it may be only imperfectly achieved, this harmonization functions to maintain the legitimacy of capitalist social relations, social order, cohesion and the authority of the state itself. This is effected through ideological or educational programmes and is ultimately enforced through the legal system and the state's monopoly of legitimate violence.

Although representing concepts and relationships diagrammatically can be misleading, depicting them as 'things', Figure 3.2 nonetheless provides a useful device for understanding Marxist interpretations of the relationship between the economic base and the political and ideological superstructure (COBI, n.d., 52). The 'relations of production' change with societal evolution, as does the manner in which surplus product is 'pumped out' of the direct producers. In classical antiquity this was by the direct expropriation of the products of slaves. Under feudalism, payment is by corvée (labour days), a proportion of production or cash rent. In capitalism, the realization of 'surplus value' is effected through the market sale of the products of labourers employed under legal conditions and ideologies that legitimate the 'free exchange' between contracting parties; employer and employee (see Pashukanis, 1924, 1978).

Historically, once privatized claims over the means of production became established, class-based society arose and the political state emerged in order to assuage the potentially inflammatory relationship between owners and non-owners, thereby securing for the 'ruling class' its control of the economic base. The Marxist perspective views the contemporary state's function as supporting the dominance of capital (ultimately a social relationship founded upon the concept of individualized private ownership right) by guaranteeing the property relations of capitalism through force if necessary (i.e. police and army). More normally, it engages in policies to encourage assent to the continuation of this dominance through ideological programmes (education, etc.). State policies are implicated in naturalizing capitalism as the only viable system of economic relations. Not only are people led to believe that capitalism is the best system

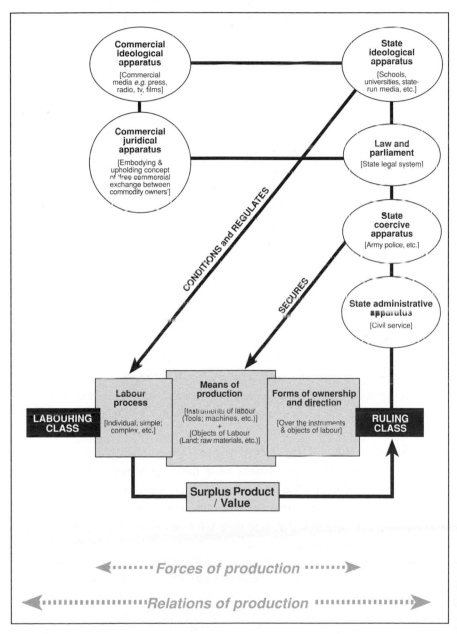

Figure 3.2 *Relationship between the economic base and the political and ideological superstructure. (After COBI, n.d., 52)*

possible, most come to regard it as the *only* viable system; a wholly legitimate and natural fact of modern life. Thus, it becomes almost impossible to conceptualize qualitatively radical change to the prevailing relations of production.

Despite its inherent relationship with the economic sphere, the state seems to act as a separate institution detached from any specific interest, appearing to be neutral between conflicting groups. Thus, when the survival of the system itself becomes threatened, state policies can be devised to ensure its continuity. Appeals to integrationist ideologies, such as the 'national interest' or the 'public good', can then be drawn upon as the logic underlying policies to bolster private accumulation. The state may even act directly against the interests of some fractions of capital (or of labour), favouring those which are perceived as more immediately fundamental to the system's survival.

Political democracies therefore provide an ideal operating context for capitalism, in which the existence of a superficial equality of personal political power (the right to vote) helps to mask the underlying intense inequality of economic power. The system even appears to be alterable, if the majority only wished! Thus, the relationship between the individual and the state seems wholly different to and separate from that between the individual and capital (see Holloway and Picciotto, 1978).

The state mediates actively in the process of capital accumulation, a process replete with crises (Harvey, 1982). It attempts to regulate aspects of the economy (e.g. regulating financial markets, managing trade relations, etc.) and often provides cheap utilities (e.g. public transport, etc.) necessary to the operation of the system but which are beyond the logic of individual capitalists to produce. It also acts as a major purchaser of products from the private sector. Born of the imperative to maintain social cohesion, it further adopts a welfare role, engaging in wider 'social reproduction' by helping to ensure the continuity of the system. Thus, it contributes to the creation of a labouring class which is healthy, possesses the necessary skills according to the changing demands of capitalist production and which stands in an appropriate relationship of dependency on capital. Thus, it engages in housing programmes, in providing health services, education and skills training.

While many dimensions of state intervention may serve to fulfil social need and effect progressive elements of reform, Marxists have been highly critical of the reformist conception of state activity. For instance, referring specifically to urban planning, Scott and Roweis (1977) have argued that the state is not the product of a reformist spirit. Neither is it an impartial referee standing outside society. Rather:

> State intervention can be understood only as a continual stream of responses to the negative and disruptive outcomes of the unresolved – and in capitalist society, the unresolvable – contradiction between privatized and decentralized decision making (as imposed by the very logic of commodity production and exchange) and collective action (as imposed by the imperative of continued social cohesion). Further, in the light of this analysis, it is now possible to assert that the specific interventionist sphere of urban planning (itself only a specialized domain of state intervention) emerges, like all State

intervention, out of a web of concrete, historically-determinate conflicts and problems embedded in the social and property relations of capitalist society generally, and out of capitalist urbanization in particular.

(Scott and Roweis, 1977, 1103)

Clearly, from a Marxist perspective, the state is not a benign institution. While marginal redistributions may be effected, as a reformist perspective would suggest, Marxists insist that the state is unable to bring about fundamental qualitative change in the relations of production because of the structural constraints on the state's activities *within* capitalism. Indeed, despite its role in occasioning improvements to working-class living conditions, the historic degeneration of working-class politics into demands-on-the-state reformism might be viewed as having brought about a reduction in working-class self-reliance and a diminution of assertiveness, militancy and aggression through the effective co-option of protest. The working class thereby becomes reduced to the position of supplicant to the capitalist state. Meanwhile, conflict tends to become directed towards the state and deflected away from capital. From a Marxist perspective, capitalism and its supporting state apparatus are therefore regarded as part of a closed political-economic system with no capacity for transformation from within (Gleeson and Low, 2000).

3.6.4.1 The Marxist perspective on urbanization and planning

Marxists view the urban realm and the capitalist urbanization process as primarily comprising a domain within which the process of capital accumulation takes place (Harvey, 1985). It is a spatial form which accelerates the rotation of capital by reducing the indirect costs of production, circulation and consumption (Kirk, 1980). Fundamentally, as was pointed out in the previous chapter, the capitalist city exists to accommodate those forms of capital which are advanced in industrial, commercial and financial spheres. It also houses the labour force in its class-differentiated residential districts where labour power seeks to recuperate from the exertions of its daily involvement in the labour process, reproducing itself both physically and ideologically from day to day and from one generation to another.

This view of the urban realm as being primarily shaped by the economic and social flows demanded by capital accumulation produces a very specific understanding of the role of urban planning within capitalism. While the creation of the built environment creates opportunities for generating profit through urban development, the contradictions inherent in the capitalist economic system become reflected in that built environment. The fragmented nature of land ownership and the short-sighted perspective of property developers seeking immediate returns, create an urban landscape which is replete with environmental dislocations. In the absence of planning, because urban development under capitalism is under the control of numerous competitive private landowners and developers, the final outcome of the land development process is in the longer-term fundamentally out of control (see Scott,

1980). In order to minimize the consequence of the myopic nature of private decision-making, planning operates as a means by which the state attempts to regulate urban land use, its development and redevelopment. Thus, Scott and Roweis (1977, 1103) have pointed out that:

> ... the State is compelled to intervene in various ways, and to secure the smooth continuation of capitalist society by taking initiatives that capitalist civil society...is incapable of taking for itself. Hence, the capitalist State functions as a general palliative and steering device, treating practical problems that are immune to resolution via the spontaneous (and chaotic) rationality of capitalist civil society, and leading society onward into collectively rational options consistent with existing social and property relations.

Sutcliffe (1981, 126) has also observed that 'the synchronization and general similarities of the worldwide planning movement are such as to suggest that town planning had a particular role to play in capitalist societies at a certain stage in their industrial development'. Major problems were being created by the anarchic quality of unfettered capitalist development and the built environment began to manifest the inherent contradictions of capitalist society generated by the production process (Beauregard, 1996; Scott and Roweis, 1977). Planning therefore emerged in response to the need for some collective mechanism to help to control and guide urban land markets towards orderly and efficient development.

So, from the Marxist perspective, urban planning is viewed as facilitating capitalism in general through attempting to create a landscape which reduces locational contradictions, enhances economic efficiency and social compliance. Planners are therefore regarded as state agents, ultimately serving capitalist interests. Thus, Knox and Cullen (1981, 184) have suggested that 'planning ... was born a hybrid creature, dedicated on the one hand to humanistic reform, but charged on the other with the management of urban land and services according to the imperatives of a particular mode of production'. They interpret it as an 'internal survival mechanism' of the capitalist economy, now institutionalized and directed by the logic of the capitalist system, which produces a series of imperatives to which planners are obliged to accord.

Scott and Roweis (1977, 1106–7) have further argued that because the structural causes of problems cannot be addressed, planning is a never-ending round of palliative and piecemeal measures because urban planning:

> ... tends to be restricted to an after-the-fact search for 'feasible' remedies to the negative outcomes of this contradictory process of urban land development. The social and property relations that underpin the capitalist land-development process are, in actual planning operations, taken to be immutable constraints. As a result, urban planning activities themselves get reduced in practice to the status of simple additional components of the whole anarchical land-development process.

They also affirm that 'the capitalist State does not have and can never have a mandate to change fundamental capitalist social and property relations, it can only ever modify the parameters but not the intrinsic logic of the whole urban land development process' (Scott and Roweis, 1977, 1110). From a Marxist perspective, urban planning thus becomes both necessary yet impossible (Fogelsong, 1986).

As a function of the capitalist state, it is clear that urban planning can never be politically neutral. Its inherently political nature was recognized by advocacy planners in the 1960s. They pointed out that the values and interests pursed in planning differed between interest groups and rejected the consensual and technical views of planning in favour of a political view. Equity planners also showed that many so-called consensual planning decisions, enacted in the name of the 'common good', tended instead to reflect and support the interests of a social élite.

Although planning theory also recognizes the political nature of planning activities (Krumholz, 1994), the Marxist perspective views this struggle as inexorably biased towards securing the interests of capital. It regards the longevity of concepts such as the 'common good' or 'proper planning and development' as a reflection of a fundamental failure to recognize that such notions discriminate in favour of certain interests, primarily of a business and property nature, as they are based on a belief that what is good for capitalism is good for society in general (Kirk, 1980).

Moreover, Marxists point out that planning goals embodied in policies and development plans are commonly dependent on the logic of the private sector for their fulfilment because they rely heavily for their implementation upon the activities of private developers. This dependency on capitalist development interests reflects the structural power of private capital, ensuring that planning policies and development plans must be drawn up to be complementary to market processes. Plans which run counter to profitability criteria, be they otherwise highly desirable, become merely aspirational if public-sector funds are not available for their implementation.

Thus, irrespective of any formal channels for public participation in urban planning, it is the ownership of private property, conferring the power to withdraw from the development process, which ensures that property interests must be favoured. It is as though planning policies, often embodied in formal development plans and related documentation, are 'ghost written' by an invisible hand which serves property interests (McGuirk, 1995). Planners' actions are therefore implicitly political as they inherently favour some interests, notably property and business interests, over others. Thus, a planning procedure such as zoning provides a useful framework within which the development industry can operate, reducing the risk to property investments of undesirable spill-over effects caused by neighbouring incompatible land uses.

Indeed, planners themselves often possess an implicit belief in a co-operative model of society and planning policies frequently reflect the idea that society comprises groups with shared beliefs, values and aspirations and that co-operation to the mutual advantage of all is normal. Thus, planning policies might claim to facilitate the development of 'socially balanced and integrated communities' in order to maximize 'economic and social efficiency'. Planning literature and planners' documents also tend to be replete with phrases such as 'orderly development', 'proper planning', 'the common good' and 'the public interest', without these ever being satisfactorily specified.

Planning takes on an entirely different significance if society is actually characterized by conflict between groups that do not share the same values or aspirations and which are identifiable by their different levels of resources, power and status (Kirk, 1980). Instead of its simply being a question of how generally agreed societal objectives might be achieved, planning becomes a matter of why certain objectives should be pursued rather than others. Thus, because planning acts as a distributor of material well-being, it is imbued with political significance. Rather than being technical, neutral, problem-solvers acting in the 'public interest', characterized by rationality and political neutrality, insofar as they accept one ideology rather than another, planners represent and act in the interests of certain social groups and classes rather than others.

Furthermore, Marxist observers have claimed that the very practice of urban planning in capitalism is diversionary as it mystifies the causes of urban problems by falsely claiming that solutions are to be found in the remedies which planners propose (Castells, 1978). Such critics have contradicted the idea that planning operates in the 'public interest', insisting instead that class interests are the driving force (Sandercock, 1998). As a result, just as with the capitalist state itself, planning cannot be viewed as a progressive force as it is deeply implicated in the reproduction of inequitable social and economic relations.

3.6.5 Post-Marxian critiques and postmodern planning

Marxist interpretations of the role of the state and of urban planning have themselves been subjected to criticism, initially from feminist and more recently from postmodern planning theorists. Though both Marxist and postmodern theorists aim to address radical inequalities in power, opportunities and resources, postmodern theorists point out that the Marxist perspective prioritizes class analysis, adopting class as the defining axis of social cleavage and the basis of societal conflict. Critics have argued that this focus on class provides a one-dimensional view of conflict and cleavage (Healey, 1996). The tendency to focus on economic struggles as the primary expression of unequal relations of power in society has tended to result in the dismissal of non-economic and non-class issues as unimportant or, at very least, their being subsumed into a class analysis. This, they argue, leads to ignoring those forms of domination which are unrelated to class domination.

Post-Marxian perspectives draw attention to diverse forms of oppression, exploitation, domination and disadvantage that relate not just to class but also to gender, ethnicity, sexuality and various forms of cultural difference. They highlight the diversity that exists within communities which, although they might be identified as belonging to a particular social class, additionally contain a range of conflicting interests and axes of division. These also contribute to shaping the growth of cities and the form and meaning of urban built environments. Identifying groups economically, by class, and ascribing them a social identity on this basis is viewed as presenting an overly-simplified picture (Gibson, 1998). Within such groups, oppression might be experienced based on differences other than class, such as racism, sexism or homophobia for instance (Sandercock, 1998).

From a post-Marxian perspective, analysis focusing on class alone is viewed as insensitive to diversity within class-defined communities and has viewed cultural-, ethnic- or gender-based oppression as necessarily secondary to, and in any case subsumed within, class-based oppression regardless of how it is experienced by individuals. Postmodernists claim that to address these oppressions and, hence, their expression in and through the built environment, would require more than addressing economic inequalities or their basis in capitalist class relations.

These critiques have led to prescriptions for postmodern urban planning; a form of planning that rejects modernist planning's assumption both of a common experience of the built environment and that of a rational, objective basis for understanding and, ultimately, controlling it. Rather than assuming equality between individuals, or any fundamental convergence in social values, postmodern planning recognizes difference (Milroy-Moore, 1996). It accepts that individuals experience and encounter the urban environment differently, depending on their class, age, gender, sexuality, cultural and moral value systems.

For postmodernists, the growing socio-cultural diversity of cities and the increasing complexity of class divisions in the global economy confirmed that it was impossible to claim a single set of 'common interests' across all socio-cultural groups in whose name planning could be conducted. Certainly, class alone could not be assumed to provide a basis for common interest. Nor could planning decisions be claimed to generate universally advantageous outcomes (Fincher, 1998). Instead, planners make assumptions about culturally-expected behaviour and regulate the built environment to accommodate it, inscribing the dominant culture spatially and thus reinforcing the status quo (Gleeson and Low, 2000) Thus, planning policies have differential and often oppressive impacts for individuals and communities depending on their complex social positioning defined with respect not only to class, but also to age, gender, sexuality and cultural values.

In extolling this complex view of social diversity, riven with multiple axes of oppression and characterized by divergent values, the function of planning for postmodernists becomes qualitatively different. It expands far beyond its traditional role of attempting to control the production and use of space and the management of land and property rights conforming with either a notion of 'the orderly city' or, as Marxists would argue, with the logic and imperatives of capitalism. Instead, to paraphrase Healey (1997), urban planning becomes a diverse practice of 'shaping places', managing collective concerns about the qualities of shared spaces and urban environments and attempting to capture and respond to the diversity of ways of experiencing, encountering and valuing the urban environment.

In particular, postmodern planning concerns the use of the planning process to help empower those whose interests are systematically ignored by conventional modernist planning practice. It abandons the modernist assumptions of rationality, comprehensiveness, objectivity and the public interest, and becomes explicitly value-driven, identifying the social values being pursued and the various social, cultural and economic interests being served in any planning instance. Furthermore, it attempts to create a built environment that can accommodate these diverse interests instead of serving dominant economic and cultural interests as has hitherto been the practice of conventional planning.

Critics of postmodernism have argued that the postmodern tendency to celebrate diversity is problematic. First it can be indiscriminate, celebrating diversity for its own sake and removing it from the political-economic context and relations in which it is created. The celebration of social diversity, much of which would be regarded by political economists as simply the expression of economic disadvantage, could therefore be used in justifying, for instance, the social diversity arising from growing socio-economic polarization (see Gleeson and Low, 2000). The emphasis on diversity can also lead to a loss of a 'progressive' ethos; the urge to achieve improvement and social reform. Some have claimed that the postmodern emphasis on diversity presents such a fragmented vision of society that the possibility of collective action, or indeed of concerted political action, is dissipated (Goodchild, 1990). Such political paralysis can effectively further exclude marginalized social groups, rendering impossible political action targeting quantifiable improvements in the equity of wealth and resource distributions and in the material well-being of disadvantaged citizens, thereby empowering structurally privileged groups.

Debates about postmodern urban diversity have undoubtedly introduced growing complexity and ambiguity into the process of devising 'progressive' or 'reformist' public policy wrought through urban planning. Certainly there has been considerable debate within postmodernism about what would constitute a more 'just' society to which planning might aim to contribute. What, for example, might constitute a commonly-held ethical framework for evaluating planning decisions and outcomes in the name of some form of a 'public interest': a public interest defined by diversity, evolving over time and varying between different places (see Gleeson and Low, 2000). For postmodernists, such an approach might be able to address inequities and injustices that are that are based not just on class but also on gender, sexuality, religion or ethnicity.

From a Marxist perspective, however, the very language in which the postmodernist debate about planning is conducted demonstrates a tendency to overlook the intimacy of the relationship between economic forms, political practice and modes of thought; between economic base, political practice and ideological superstructures. Addressing inequities and injustices emerging from the oppressive power relations of sexism, racism or imperialism can never of itself result in the transcendence of the capitalist relations of production which underlie the more fundamental inequalities in the human condition. Only the transformation of the class basis of economic power can achieve this.

3.6.5.1 Postmodern planning and the state

This inevitably brings us back to the question of the role of the state and the positioning of urban planning as a state activity. As discussed earlier in this chapter, from a Marxist perspective, the state is viewed as intrinsically linked to the economic sphere, the instrument of the capitalist class in its role as guarantor of the necessary relations and conditions for capital accumulation and as a suppressor of potentially threatening class conflict.

Although the postmodern perspective accepts the structural basis of inequalities of power between groups, it does not view these as being rooted entirely in the mode of production and the class-basis of capitalism, nor as necessarily being supported by the state. Instead, power inequalities are defined by more than economic or class status. Thus, it maintains that a more 'equitable' society is to be achieved not only through redistribution but also through the recognition of differences in values and cultural expression. This means planning for cultural diversity: recognizing the diversity in which the urban environment is experienced and attempting to accommodate a diversity of ways of living within it.

However, the postmodern perspective views the political sphere of the state and the economic realm as being separable, even if they are often related. While it accepts that the state may be implicated in supporting the *status quo* regarding the balance of class and cultural power, it maintains that greater 'social justice' can be achieved through state based urban planning.

Thus, the postmodern perspective maintains a fundamentally reformist and essentially pluralist vision of the state. The state is not viewed primarily as the guarantor of capitalist social relations, therefore it is not positioned as an adversary with a pre-determined antagonism towards interests beyond those of capital. Rather, this perspective envisages 'a complementary ... relationship between the state and civil society and ... the possibility of social transformation as a result of the impacts on the state of mobilized civil society' (Sandercock, 1998, 102). In other words, urban planning which involves the culturally-sensitive inclusion of a range of social groups can be a route towards progressive reform, working towards outcomes which offer recognition to and address the needs, interests and values (economic and otherwise) of disempowered groups.

However, it is also argued that planners need to seek other forms of empowerment and recognition for those disadvantaged by structural inequalities, by working outside the confines of their role as functionaries of the state and beyond the modernist assumptions in which its institutions and bureaucracies are steeped. In a manner highly reminiscent of 'advocacy' and 'equity planning', planners are required to adopt roles which involve engaging in the actions of local community organizations helping to increase their social and political power (Sandercock, 1998).

3.7 Urban planning and urban development

As the preceding sections demonstrate, the interpretation which one adopts concerning the role of the state in capitalist society will inevitably determine one's conception of the role of planning. Planners may be viewed as technical operatives implementing democratic decisions, as managers and bureaucrats with ideals and goals of their own making, as social reformers effecting marginal redistributions in favour of the less well off, or as functionaries of the local state whose actions favour dominant classes or interests and provide legitimacy to outcomes which are intensely unequal (Kirk, 1980; Knox, 1982; McGuirk, 1991). More recently, postmodernists might view them as place managers, operating both from within and outside the state to 'shape places' in line with diverse cultural, social and economic values. There is some validity in each perspective, though Marxist interpretations clearly offer the most penetrating analyses.

Specifically, urban planning in capitalism addresses those land-use and development issues which result from the fragmented character of property ownership and which derive from the myopic decision-making of private owners interested primarily in the immediate profitability of their own development scheme. However, urban land cannot be 'consumed' in the same manner as other commodities. Urban land is fixed in space. While it has an absolute position, it also possesses a 'relative location'. Its successful development and use are therefore replete with locational interdependence. The profitability of any development is therefore highly dependent on what happens on adjacent sites, because 'inappropriate' development can create negative externality effects causing the devaluation of adjacent properties. Scott (1980) also highlights the existence of bottlenecks in land development arising from site assembly difficulties, where owners of key sites consider the economic life of their properties to have not yet passed their maturity and refuse to consider redevelopment. He further points to problems of 'free riding'. For example, successful land development may require the provision of certain infrastructural items such as trunk drainage or a costly bridge to create road access. Once provided, its use by neighbouring landowners enables them to 'free ride' on the original developer's investment. Similarly, in areas where profitability from redevelopment is uncertain, it pays site owners to delay development until others have borne the pioneering risk. Postponement permits the 'free rider' to benefit from a safer development context. But as all site owners are making the same calculations, each may decide to wait for others to go ahead, engendering inactivity and the deterioration of buildings.

These land development problems are addressed through urban planning. Land-use zoning prevents the development of incompatible land uses. The state's powers of compulsory purchase (eminent domain) can overcome bottlenecks in land development by facilitating site acqusition and assembly. Finally, problems of 'free riding' are addressed by state provision of infrastructure, paid for out of general taxation or from special levies so that none can 'escape' payment. Or the state may provide fiscal incentives to promote redevelopment in risky locations.

Scott (1980) also identified several 'dynamic' land-use problems. One example derives from the slow convertibility of land. Because of the enormous investment embodied in buildings, urban land uses can respond only slowly to changing patterns of demand. Thus, non-optimal locations and inefficiencies in land-use arrangements tend to become 'locked' into the urban landscape. This can be seen where industrial operations are rendered sub-optimal by encirling urban growth. Relocation might be a possible response but is not easily effected because of the costly fixed-capital investment involved. Second, significant differences can arise between the optimal timing of development from the perspective of the private developer and a broader societal viewpoint. Thus, planners may restrict development at certain times in some locations by refusing permission to develop or by not providing necessary infrastructure, such as mains drainage, in order to delay development. Elsewhere, or at other times, the state may promote development by offering advantageous planning permissions or fiscal incentives. Third, because private-sector development decisions tend to be made with immediate profitability considerations in mind, this temporal myopia tends to drive spatial arrangements towards outcomes which may be inefficient and ultimately irrational in the long term. Urban planning provides a framework for guiding longer-term private-sector decision making towards broader strategic spatial outcomes.

As was noted earlier, it is precisely because urban land development is privately controlled that, in the absence of planning, the final outcomes of this process would be fundamentally out of control (Scott, 1980). Yet, the creation of an efficient spatial system is imperative for the successful functioning of the economic system. Collective action by part of the state is therefore required. Urban planning intervenes in the process of land development, or redevelopment, in order to limit the degree to which the individual short-term perspectives of the multitude of private-sector developers are able collectively to drive the aggregate outcome of urban development towards spatial arrangements which are inefficient.

However, as is true of a great deal of state action within capitalism, urban planning's interventions often serve merely to displace problems, resulting in their later reappearance in another form, perhaps also in a different location. This gives rise to a sequence of problem-oriented responses to continuing manifestations of underlying and unresolvable contradictions within the space-economy of capitalism.

3.8 Planners and developers

It was noted above that collective action on the part of the state, operating through urban planning, facilitates the successful outcome of property development in capitalist cities. Despite their mutual dependence, the relationship between planners and developers can nevertheless be characterized as having been traditionally one of mutual suspicion, if not one of outright hostility. Certainly, there is a tendency for planners and developers to have very different opinions regarding the value of planning. As was discussed previously, planners have traditionally had a tendency to view themselves as guardians of the 'public interest'. Peiser (1990, 497) caricatures their self-image in a purposely exaggerated way:

> They represent the public interest, the generations yet unborn. They give political voice to the poor and the powerless. They provide a balance of power to the rich and influential (developers). In the planning and public decision-making process, planners are the neutral analysts and arbiters, the voice of disinterested reason among political factions. Most planners have a mission to create a just and more humane society.

As 'evangelistic bureaucrats' (Davies, 1972) with a reformist mission, a caricatured version of planners' views of developers shows them to be suspicious of developers' profit motivation, believing them to care little for community needs and possessing little interest in producing long-term equitable urban outcomes or resource allocations.

In turn, developers frequently regard urban planning as a costly and unnecessary bureaucratic hurdle that has to be overcome in what they view as their entirely legitimate pursuit of development profits. Planners are therefore often perceived as being negative, overtly obstructive and irresponsible. Developers have cast them as 'leftist pinkos' and 'closet commies' (see McGuirk, 1991) unwilling to understand the highly risky business of creating the built environment. While 'undue' planning delays which planners are believed to cause can be critical in determining whether a scheme generates profit or results in loss (Peiser, 1990), they seem ignorant that appropriate development timing is crucial for profitable development. To many developers, planners seem altogether unsympathetic to the very notion of profiting from urban development.

There is some validity in both caricatures. However, because control over land and financial resources lies with developers, with the power simply to withdraw should their criteria for engagement in development not be met, the greater power lies with the development sector rather than with planners. As Peiser (1990, 498) has noted:

> Planners can tell developers what they cannot do, and suggest what they should do, but the developer makes the final decision concerning what will happen or not happen.

Indeed, it has been in response to the traditionally negative or permissory role of urban planning and its particular powerlessness at times of development inactivity that entrepreneurialism has increasingly infused the operations of contemporary planning. New and more flexible forms of engagement between planners and developers have therefore increasingly been adopted, involving joint ventures or fiscal incentives.

Moreover, urban planners are obliged to respond to a wide diversity of social groups, ranging from the demands of property and business interests to those of residential communities differentiated by class, lifestyle and ethnic identities whose demands on urban planning may be quite distinct and a source of potential conflict. Peiser (1990, 499) further acknowledges that the immediate, complex context within which planners have to operate is a major constraint on their power to influence outcomes, as they have to function in a number of contexts simultaneously.

> Planners have difficulty asserting themselves effectively because they serve in a variety of conflicting roles and represent different constituencies. They function in four sometimes conflicting positions: staff member, analyst, arbiter, and advocate. Similarly, they serve four divergent constituencies: elected and appointed city officials; citizens, including residents, workers, and employers; implementers, both private and public developers and a variety of public agencies engaged in infrastructure development; and, last but not least, the public interest, generations yet unborn.

Peiser (1990, 498) has further claimed that

> if planners properly understood the role of risk taking and profit in development, and if developers understood that most planners are as eager as they are to generate high quality development, planners and developers would be able to work together better and accomplish mutually shared goals faster.

However, the earlier sections of this chapter have suggested that there exist broader structural constraints upon planning, born of the over-arching limitations set by the role of the state itself within capitalism. Moreover, during the later years of the twentieth century, the character of many urban planning systems was to become increasingly transformed as the traditionally restrictive and sometimes conflictual stance of planning with respect to development interests shifted towards a more overtly facilitative position towards the private sector.

To a degree, this was an almost inevitable consequence of the declining power of traditional urban planning after the 1960s. In the housing development context in the UK, Ball (1983, 246) showed that the real power of urban planning had 'centred around the ability of the state to orchestrate land use development through large-scale public expenditure on the built environment'. Public-sector spending on infrastructure had allowed planning authorities effectively to manipulate the urban profitability surface of housing developers by providing for higher development gains in certain locations, where planners wanted development to happen, rather than at others. However, this power to influence developers' decisions was undercut by reductions in public spending associated with the deepening fiscal difficulties of the state during the 1970s.

The reduction in planning's power was especially evident during a period of economic stagnation when the number of development proposals also diminished. By the end of the 1970s, Goldsmith (1980, 144) was led to comment that 'planners now appear as rather weak, perhaps irrelevant, pawns in an economic and political environment which is much more hostile than it was twenty years ago'. Urban planning had been rendered an almost inconsequential administrative exercise and, clearly, other forms of intervention needed to be sought.

From the 1980s, that search became embodied in the evolution of a deepening entrepreneurialism. Urban planning sought to work together with the forces of property development to bring about planning goals by ensuring that plans conformed to developers' profitability considerations. Where development risk proved too great, 'pump-priming' public spending would create necessary infrastructure or fiscal inducements would be provided to subsidize development costs. Moreover, planning increasingly became part of a strategic process of urban development, providing a guiding role in urban development in response to the agendas of urban 'boosterism' lobbies. The case study of Sioux Falls shows clearly how a strategic alliance was forged between urban planning, the economic growth lobby and the forces of property development, re-inventing both the image and reality of the city by helping to create an urban landscape which was iconographic of economic success, growth and change.

3.9 Entrepreneurialism in urban planning

Active public-sector involvement in urban development is longstanding. The provision of social housing by local authorities in the UK, for example, can be traced back into the first half of the nineteenth century. On a broader scale, responsibility for the creation of completely new towns in the UK after World War 2 was given to public-sector special purpose agencies, the new town development corporations (see Schaffer, 1972; Osborn and Whittick, 1977). With legal and financial powers to acquire and develop land, they provided everything from roads, water and sewerage infrastructure, to housing for a wide range of social classes, in addition to developing factories, shops and offices. Public–private partnerships had also been used frequently to facilitate city-centre redevelopment in the UK during the 1960s. Typically, local authorities would use their powers of compulsory purchase to assemble sites. These would then be developed by private-sector developers, using their development expertise and finance-raising powers, the local authority benefiting from an element of 'planning gain' in the form of some community facility such as a library or social centre.

In the USA, the initiative for promoting urban redevelopment has often originated with a local coalition of business leaders and government officials (Fainstein, 1994). Renewal policies pursued in many US cities since the 1949 Housing Act saw city governments 'acting effectively as brokers for private developers' (Kivell, 1993, 8), assembling sites for redevelopment and providing necessary infrastructure. Even the 1977 Housing and Urban Development Act which made provision for Urban Development Action Grants (UDAG) to tackle poverty and housing problems were instead often diverted into subsidizing private development benefiting property developers and the more affluent (Feagin, 1983; Fainstein, 1994). These entrepreneurial aspects of American urban planning are explored in some depth in the case studies of Minneapolis and Sioux Falls.

However, the last two decades of the twentieth century were marked by an increasing degree of entrepreneurialism in urban planning which was often of a markedly different character to previous public–private partnerships. Earlier examples had commonly reflected a reformist planning ideology and a response to the necessity to reconfigure urban environments to facilitate the changing demands of capitalist operations. Latterly, change stemmed from a growing entrepreneurial culture in urban governance itself, largely directed and orchestrated by central-government initiatives (McGuirk, 1994).

Under 'Thatcherism' in the UK, 'Reaganomics' in the USA and similar programmes grounded in the politics of New Liberalism in New Zealand and Australia, the role of the state was to become more overtly geared towards serving the interests of capital, presiding over major shifts in the distribution of the social product away from labour power and towards capital. Paradoxically, contrary to the anti-public-sector rhetoric of the neo-liberal agenda, the state devoted massive public subsidies to the task of attracting capital back to the inner city: the 'feeding of the downtown monster' (Harvey, 2000, 141). The state was also increasingly obliged to deepen its social containment role in order to repress the industrial unrest and political revolt resulting from the destructive social consequences of neo-liberal free-market utopianism and the politically naïve faith in its inevitability (see Harvey, 2000).

Entrepreneurial styles of governance brought changes in the methods and instruments of regulation. It generated a re-distribution of responsibilities across the different spheres of government, the community and the private sectors, forcing a transformation in the cultures of interaction between them (Painter, 1997). It became associated with networks of negotiation linking government departments, quasi government organizations, private sector and voluntary-sector interests. The new governance regimes engendered a 'creeping enfeeblement' of local government, typified by public–private partnerships, appointed quangos, alliances with non-government actors and entrepreneurial initiatives (Peck and Tickell, 1994).

Simultaneously, during the 1980s, property-based urban regeneration strategies became an increasingly widely used tool of urban renewal policies (Turok, 1992; Williams and MacLaran, 1996; Newman and Thornley, 1997; McGuirk and MacLaran, 2001). A range of different strategies marked the emergence of an entrepreneurial culture around urban regeneration (Deakin and Edwards, 1993). Common features of this new entrepreneurialism included the use of Special-Purpose Authorities (SPAs), property-led renewal initiatives (Healey et al., 1992), flagship developments (Smyth, 1994) and revitalization based on the exploitation of cultural capital (Boyle and Hughes, 1994).

In this new configuration of governance, the context for urban planning became transformed. Local government planning powers were sometimes rescinded altogether and vested in alternative, often centralized, quasi-private agencies, thereby weakening local input or control. Indeed, planning functions were frequently scattered across a range of area-based development agencies to the extent that an absence of strategic spatial planning might arise (see Newman and Thornley, 1997). Furthermore, the adoption of entrepreneurial forms of governance heightened the necessity of complying with the aims of economic and financial interests especially where area-based special-purpose authorities were created to oversee redevelopment (Malone, 1996).

For example, the remit of the London Docklands Development Corporation (LDDC) was to bring land into 'productive' use. Within its geographical area of competence, planning powers were expropriated from the docklands local authorities and vested in the LDDC. The corporation overtly subsidized property development through enormous investment in infrastructure to improve the accessibility of docklands. It used its powers to assemble land, compulsorily if necessary, to treat it if for industrial pollution and to convey it to private-sector developers at much reduced real cost (see Ambrose, 1986; Brownill, 1990; Fainstein, 1994).

Entrepreneurial planning sometimes became focused around a narrowly conceived goal of physical renewal driven by financial incentives (McGuirk and MacLaran, 2001). Alternatively, more active forms of partnership extended beyond simple physical regeneration into broader economic development, education and training, and the provision of entrepreneurial leadership (Peck, 1995). Operating under such conditions, urban planning can sometimes appear to become a depoliticized form of development planning locked into a pro-growth agenda, as the Sioux Falls case study shows. In such cases, broader social, cultural or environmental objectives can become residualized or diluted, overridden by a centralized agenda of competition, growth and entrepreneurialism (Imrie and Thomas, 1995).

In planning systems characterized by a separation into development planning (e.g. research, policy formulation and the writing of development plans) and more mundane functions of development control (i.e. negotiations with developers over prospective developments) another consequence can be the reinforcement of existing divisions between the formal planning system and those social issues embodied in other government programmes (Newman and Thornley, 1996).

Figure 3.3 *The London Docklands Development Corporation acted as catalyst opening up docklands to private-sector property development and investment*

Figure 3.4 *Fiscal incentives have frequently been used to influence the profitability surface for development*

Entrepreneurial planning, with its common reliance on special purpose authorities, is highly centralized and reliant on a fluid mode of operation in which boundaries between the public, private and community sectors become blurred. Its functioning relies on networks of negotiation between these sectors rather than on hierarchically determined bureaucracies (Rhodes, 1988). Thus, despite the tendency for local governments to become disempowered by entrepreneurial systems of urban governance, it has been suggested that the new institutional forms and channels of communication present opportunities which might be exploited to the advantage of local governments (see Healey *et al.*, 1995; Bassett, 1996).

It has been proposed that well-directed opportunism could allow planners to become more central to the process of policy making and plan implementation (Stoker and Young, 1993). These capacities might enable local-government planners to mobilize effective networks that could work in an integrated fashion towards achieving broad environmental, economic, social and cultural planning aims. Stoker and Young (1993) have argued that local government and local government planners could become important contributors to tackling urban problems if they could develop a new style of operation amenable to contemporary modes of governance. This would require a departure from hierarchical and bureaucratically-determined practices driven by rules and regulations, which are slow to respond to new demands arising from altered policy and socio-economic environments.

While such a shift in operations may seem to offer the opportunity to become more closely involved in key decision-making processes, this potential may be rather more apparent than real. As we have already seen, because most development activity is led by the private sector and depends on profitability criteria, such a reliance inevitably emphasizes the imperative of maintaining a corporate ethos and modes of planning practice which are supportive of property capital (Leitner and Garner, 1993; Wilks-Heeg, 1996). This is unlikely to be a recipe for any radical change that could prove detrimental to the interests of capital.

3.10 Conclusions

This chapter has reviewed the ideological foundations of contemporary planning practice and addressed recent changes in its operations to accommodate more entrepreneurial styles of engagement with the private sector. It has also presented a number of different perspectives on the nature of the state and why urban planning has sought to intervene in the process of urban development. Moreover, it has shown that, as a function of the capitalist state, the potential of urban planning to generate outcomes which are detrimental to the interests of capital are clearly structurally constrained.

As systems of entrepreneurial urban governance become increasingly prevalent internationally, so the organization and direction of urban planning systems have tended to mimic those of American cities, which have long been more overtly facilitative of private-sector interests. This facilitative mode of planning challenges the professed neutrality of urban planning, exposing the alignment of the state with private development interests. In the following case-study chapters, many of these themes are taken up in greater depth, exploring aspects of the relationship between urban planning and the property development sector in different urban contexts.

4

The rejuvenation of downtown Minneapolis: urban planning as a creature of private-sector interests

Andrew MacLaran and David Laverny-Rafter

4.1 Introduction

This chapter examines the rejuvenation of Minneapolis's central business area since the 1960s. It reviews the market forces which drove the city's property-owning and business élite to address the problems of central-area decline, first through the co-ordination of private initiatives and second, by persuading the city council to create a fully-fledged urban planning system as a means of securing their interests. It describes the intimacy of the relationship between private-sector interests and initiatives and the role of public-sector policies that facilitated the successful transformation of the downtown. It demonstrates above all that urban planning was a creature of private-sector demands, established to address serious problems threatening the viability of downtown businesses and real estate values.

The difficulties facing downtown Minneapolis resulted from rapid suburbanization, a changing economic base and uncertainty associated with new infrastructural developments. In setting its downtown goals, planning embraced the aims of central-city property owners and business sectors by embodying them in planning strategies and supporting them with financial inducements and a business-friendly environment for property development. Only in the 1990s did critical evaluation of these overtly pro-business planning policies arise, bringing about some redirection of development priorities by trying to provide a greater degree of benefit for the wider community. However, powerful political forces continue to press for urban planning to revert to its overtly business-supportive role.

4.2 The context

By the late 1950s, Minneapolis was experiencing a decline in the economic activity of its central business area as a result of peripheral residential expansion and competition from the expanding suburban economy, which had one of the fastest employment growth rates of any American city. During the 1960s, the suburban population grew by over 55 per cent while the central-city population fell by 6.5 per cent. Central-area employment rose by less than 2 per cent while employment in suburbia increased by 120 per cent

(Reuss, 1977). These trends had major consequences for central-area property owners, a feature already observed in St Paul prior to World War 2. One report noted with concern the:

> firmly-established trend toward de-centralization in Saint Paul, and the ruinous effect it is having upon downtown commercial property. This has already caused staggering losses in the real value of real estate in the city's heart.
>
> (Organizing Committee of the City of St Paul Central
> Business District Authority, 1943, 3)

Although downtown Minneapolis remained the commercial and economic focus of the Upper Midwest during the 1950s (Aschman, 1971), ominous trends pointed towards decline. Residential suburbanization was followed by the development of the first suburban shopping centre in 1939. By 1958 there were 57 such centres in the Twin Cities (Gilhousen, 1985). The largest, Southdale Shopping Center, developed by the Dayton Company in suburban Edina, was the first multi-level enclosed mall in the USA. It was opened in 1956 and totalled 75,000 sq. m. (800,000 sq. ft.). Occupiers included both Dayton's and Donaldson's, the leading department stores on downtown Nicollet Avenue, Minneapolis's prime retailing street. The trend represented a clear threat to downtown retailing. Between 1958 and 1963, 48 retail establishments left central Minneapolis, mainly from the edge of the central business district, and downtown retail sales fell by 10 per cent. Additionally, the old industrial areas in the North Loop on the fringe of downtown were in decline. Over 50 businesses left central Minneapolis between 1945 and 1970 (Minneapolis Planning and Development, 1970). The decision in 1954 by General Mills to move to a new headquarters in the suburb of Golden Valley, had major psychological impact.

Given this typically American pattern of downtown economic decline, it has been a significant accomplishment that, within 25 years, downtown Minneapolis was to become reinvigorated economically and transformed physically, with widespread rehabilitation and upgrading of older buildings and the development of major commercial property schemes. The 1970s and 1980s saw the construction of over 45 major developments each valued at over $10 million, more than 10 of which exceeded $50 million. Developments included retail and office space, hotel accommodation, residential units, multi-storey parking ramps and a convention centre. These were the product of both private- and public-sector initiatives, though it was the business sector which acted as the catalyst.

4.3 Private sector initiatives

4.3.1 The Downtown Council and the reconstitution of urban planning

Faltering attempts to establish a planning system in Minneapolis had existed prior to the 1950s. As part of the 'city beautiful movement' a Civic Commission was formed in 1910 and given the task of drawing up a master plan for the city. Although this was published in 1917, the plan had few tangible consequences. In 1921, the first City Planning Commission was established with the power to prepare a city plan, propose public improvements and recommend zoning ordinances. Implementation of the 1924 plan addressed the problem of growing automobile use, but was hampered by the Depression. A second comprehensive street plan, published in 1940, was aborted by World War 2. However, once powerful private-sector interests became threatened with the devaluation of their central-area capital investments, public intervention became doggedly pursued.

Three factors highlighted the need for a more fully fledged planning system. First, by the mid-1950s it was obvious that the 1924 zoning ordinance had become outdated. Second, in 1949, the city embarked on massive urban renewal in the Gateway district of declining and abandoned industrial property in the North Loop. Funded to the extent of 75 per cent from federal sources and administered by the Minneapolis HRA (Housing and Redevelopment Authority), this area of derelict and abandoned properties, grain and lumber mills, 'cage hotels' of lofts subdivided by chicken-wire and other low-grade residential accommodation was rapidly cleared. Third, there had been no detailed evaluation of the impact on downtown Minneapolis of the inter-state highway program which focused on the city centre (Figure 4.1).

Against this background of metropolitan expansion, freeway construction, industrial abandonment and the likelihood of further reductions in retail turnover downtown

Figure 4.1 *Focusing on the downtown, interstate highway I-35W cuts a swathe through suburban Minneapolis*

causing reductions in real-estate values, there developed a growing partnership between private interests and the public sector. Downtown real-estate taxes were crucially significant in financing city government. Thus, the city council had a vested interest in preserving downtown property values. Consequently, the interests and goals of downtown businesses and property owners were regarded by the city government as benefiting 'the community'. This inextricable link between the aims of capital and public policy was to prove especially fortuitous for the owners of downtown real estate.

Concerned by contemporary trends, a voluntary association of business and professional corporations established itself in 1955 as the Minneapolis Downtown Council. Its aim was to expand, enhance and conserve the downtown area by promoting the development of shopping and business activity.[1] Its board comprised senior executive officers from the leading corporations, retailers, banks, media, building owners and the owners or managers of smaller businesses (Aschman, 1971). By the mid-1980s, it represented 420 downtown businesses and had been designated as the official representative body for the downtown neighbourhood.

This self-constituted Downtown Council was instrumental in the reconstitution of planning in Minneapolis. It commissioned a Chicago-based firm of planning consultants, Barton-Aschman Associates (BAA), to advise on the most appropriate way in which to proceed. BAA advised the Downtown Council to work closely with the public sector, which alone possessed powers of zoning and eminent domain (compulsory purchase). The economic role of downtown and the importance of its property-tax base to the municipal authority proved a persuasive tool in generating city council support for a strong central area. The Downtown Council persuaded the city council to expand its Planning Department from a single professional employee in 1956 into an effective organization.

The City Planning Commission's draft Downtown Plan of 1959 recognized that most decisions concerning the development of central Minneapolis were made by private individuals. It therefore accepted that private decision-making criteria would have to be taken into account if the goals of the draft plan were to be achieved (City of Minneapolis Planning Commission, 1959, 2). It was stated that:

> It is in the best interest of the Community, the State and the Region that Central Minneapolis remain and grow as an area of intense activity and high value.
> (City of Minneapolis Planning Commission, 1959, 4)

Three basic rubrics underlay the plans. They had to be 'physically and economically feasible', 'in accord with the desires of the community' and 'capable of adapting easily to unforeseen or uncertain changes in technology, the economy or social custom' (City of Minneapolis Planning Commission, 1959, 4). These seemingly uncontentious foundations ensured that planning would serve unquestioningly the interests of the downtown business community and property owners. The draft plan established a number of planning principles which were to remain relatively unchanged over the following three decades. These provided a clear framework within which property development decisions could be undertaken with knowledge and confidence. They included:

1. Maintaining a compact downtown with clusters of different activities within easy pedestrian access of one another so as to be mutually supportive economically.
2. Encouraging mass transit use, discouraging through traffic and ensuring that the central area was accessible by providing peripheral parking, and promoting pedestrian circulation by creating pedestrian malls and upper-level walkways.
3. Enhancing the attractiveness of downtown by emphasizing its distinctive appearance and creating places for congregation and activity.

This consistency of downtown planning policy during the following 30 years became a vital component for rejuvenation. Knowledge of planners' intentions and their commitment to preserve the role, status and property values of the downtown created the climate of confidence necessary for rejuvenation and reduced the levels of risk involved in property development projects.

Thus, while the importance of downtown taxes to city budgets induced planners to formulate plans that were acceptable to the private sector, the inherent structuring of those plans to the requirements of the private sector also ensured that there was a high probability of those plans being fulfilled. Notably, where plans did not adequately take into account the criteria of the property development industry they long failed to engender firm redevelopment proposals, notably on the west side of Hennepin Avenue (Block E) where planners had insisted on what most developers considered to be an uneconomic mixture of functions.

Figure 4.2 *Hennepin Avenue at Fifth Street (1988). Near the main retailing street (Nicollet Mall), low-rise secondary commercial functions dominate*

4.3.2 Nicollet Mall

Although the draft plan was never formally adopted by the city council, key proposals were taken up by the private sector. The first concerned the main downtown shopping street, Nicollet Avenue. In 1957, the Downtown Council had formed a Nicollet Avenue Survey Committee under the Chairmanship of L. C. Park of Baker Properties. In 1956, he had proposed the redevelopment of the blocks along Nicollet Avenue into a series of enclosed plazas. Both the committee and planners increasingly accepted that the improvement of Nicollet Avenue would strengthen downtown Minneapolis, stimulate office development and help to expand the retail market (Aschman, 1971). BAA were again commissioned to work with the permanent Nicollet Avenue Subcommittee of the Downtown Council on plans for the improvement of the street. They concluded that major refurbishment was highly desirable, especially as the avenue's retail turnover fell by a further 10 per cent between 1959 and 1964 (Heath, 1984).

The consultants' report was accepted by the Downtown Council. It then commissioned BAA to draw up a general plan, while Lawrence Halprin and Associates of San Francisco were commissioned to work on the project's design and landscaping. Approval by the city council and Hennepin County Board of Commissioners provided official sanction, the project thereafter becoming 'essentially a city endeavour with planning financed by public funds but with the Downtown Council maintaining a key collaborative role' (Aschman, 1971, 4). The $3.8 million of improvements were financed by local businesses through special assessments on properties located on or close to the new Nicollet Mall, and from federal sources including an Urban Mass Transportation Demonstration Grant and an Urban Beautification Grant.

The mall had an immediate impact. Retail sales rose 14 per cent during the year after its completion in November 1967, while its psychological effect on property owners resulted in $225 million of new construction and rehabilitation of premises by 1971.

4.3.3 The Skyway system

Another element of the draft plan adopted by the private sector involved the improvement of facilities for pedestrian movement downtown. In the early 1960s, the city council had granted a right of encroachment over the roadway to the owners of a building who wanted to construct a pedestrian bridge linking an office block and a multi-storey parking ramp on either side of Marquette Avenue. From this private-sector initiative evolved Minneapolis's 'Skyway' system, providing a means of pedestrian circulation above street level comprising bridges across streets and defined routes through private buildings, linking 49 downtown blocks by 1992 and creating a safe, climate-controlled environment comparable to a suburban shopping mall (Figure 4.3). The benefit of the system is reflected in the growing use of this upper floor for retailing and in the higher than normal rents charged for floorspace located there.

Figure 4.3 *The skyway system provides a means for pedestrian downtown circulation*

4.3.2 The IDS Center

Another major event in downtown rejuvenation resulted from an intervention by Kenneth Dayton of Dayton Hudson, who chaired the Downtown Council. He persuaded the industrial holding company, IDS, to upgrade its proposal to redevelop a building on Nicollet Mall on one corner at the city's peak land value intersection. IDS increased the scale of its development project and engaged top-ranking architects Johnson and Burgee. The 52-storey IDS Center (Figure 4.4), valued at $42 million on its completion in 1973,

Figure 4.4 *The IDS Center, comprising a 51-storey office tower, a hotel, three levels of retailing and a Crystal Court covered piazza, is a focus for the skyway system*

comprised 116,000 sq. m. (1.25 million sq. ft.) of offices, 18,500 sq. m. (200,000 sq. ft.) of retail space around a glass atrium, a 282-bed hotel and parking for 600 cars. It represented a powerful symbol of private-sector confidence in the future of the downtown economy and acted as a catalyst for further rejuvenation.

4.4 Public intervention: a pro-active approach

A number of public-sector interventions underwrote the private-sector activities, thereby demonstrating the commitment of the city government to maintain the status of downtown. First, the federally-funded clearance of the Gateway area of the North Loop eliminated large areas of declining industrial and warehouse properties. A similar but smaller-scale clearance of low-income apartment buildings to the south of the central business area took place near Loring Park. Second, the creation of a fully operational urban planning system had provided the crucial contextual framework within which private-sector decision making could be conducted in a controlled, risk-reduced environment. Third, improved accessibility was achieved through the focusing of new inter-state freeways onto the downtown with access to parking facilities adjacent to the core. A fourth factor was the provision of a subsidized and frequent mass transit bus service centred on the downtown, ensuring that it would remain accessible from all parts of the metropolis.

In 1974, covering two city blocks near City Hall and just four blocks to the east of Nicollet Mall, the $41.8 million Government Center office development and its adjacent landscaped open space were completed. The development created a high-quality environment and encouraged further private-sector redevelopment in the area (see Figure 4.5), notably the Lutheran Brotherhood building, 701 Fourth Avenue and the Lincoln Center. It clearly demonstrated the power of public investment, unfettered by the criterion of profitability, to become a market leader by creating conditions conducive to further development activity, the positive spillover effects being reaped by owners of adjacent properties.

Furthermore, the Minneapolis Community Development Agency (MCDA) which reports to the city council, played a key role by adopting a highly active approach to regeneration. It encouraged redevelopment by using its authority of eminent domain (compulsory purchase) to facilitate site assembly for approved redevelopment schemes and used its powers to raise finance through issuing bonds in order to provide financial inducements for downtown redevelopment projects. With distinctive sources of finance separate from the normal city budget, set apart from the city planning authority and shielded from the accountability of the electoral process, the MCDA had the freedom to become engaged in highly entrepreneurial activities (see Leitner, 1990). At a time of general reductions in the scale of inter-governmental revenue transfers to municipalities under the Reagan administration and their need to become more financially self-reliant, the pro-development ethos of the MCDA seemed uncontentious, highly desirable and even essential.

Figure 4.5 *Development near the Government Center. The twin 24-storey towers of the Government Center with its 107 m. (350 ft.) atrium (left), the Lutheran Brotherhood 17-storey glass wedge office (centre) and the 31-storey Lincoln Center (right)*

The MCDA gave substantial inducements to private-sector developers, providing them with low-cost finance and conveying assembled sites to them at substantial discount. In so doing, it employed extensively the facility of Tax Increment Financing (TIF).

> Tax Increment Financing is a tool used by cities to finance certain types of real estate development costs. The primary purposes of TIF are to attract private investment that will 1) redevelop blighted areas, 2) provide housing for low and moderate income individuals and families, or 3) result in increased employment opportunities and tax base.
> (Office of the Legislative Auditor, State of Minnesota, Program Evaluation Division, 1986, ix)

TIF permits a city to raise capital by issuing tax-exempt bonds to finance its operations in areas designated as Tax Increment Districts (TID). The funds are then used to purchase properties, for building clearance and to 'write down' land-values, selling sites on to developers at below their cost of acquisition. This enables the city authorities to subsidize a desired property development scheme. As the new development has a higher value than the buildings that it replaced, the additional property taxes generated by the redevelopment are used to make capital repayments and interest charges.

For example, private-sector redevelopment of an importantly-situated block of physically obsolete and under-utilized commercial properties on Nicollet Mall had long been thwarted by highly framented ownership. After designation as a TID, the city

Figure 4.6 *The City Center development comprises over 100 retail units, a department store, a hotel and a 52-storey office tower. The developer benefited from a $36 million site value write-down by the MCDA*

authorities were able to use compulsory powers of acquisition (eminent domain) to assemble the site and transfer it to the Canadian developers Oxford, having written down the site value by over $36 million. This would subsequently be recouped by the city from the higher real-estate taxes on the new development, the funds being used to service the interest charges and pay back the $50 million Tax Increment Bond. The resultant $250 million 'City Center' complex included a 26,000 sq. m. (280,000 sq. ft.) department store, 17,000 sq. m. (180,000 sq. ft.) of additional retailing space, a 51-storey 100,000 sq. m. (1.08 million sq. ft.) office tower (Figure 4.6), a 600-room hotel and a 500-space parking ramp. The property's assessed land value prior to redevelopment had been just $9.2 million whereas its value in 1985 was assessed at $93.4 million (MCDA, 1987a).

Minneapolis became an enthusiastic user of TIF, though not without controversy over the scale of inducements offered to developers and the probability that some developments would have taken place without them. By 1987, the city had 38 active TIDs with a total captured assessed value of over $310 million. It had issued over $170 million worth of Tax Increment Bonds involving an effective subsidy of around 9 per cent of total development costs (MCDA, 1987a; Leitner, 1994).

4.5 The impact of policy: evidence of rejuvenation

The physical rejuvenation of downtown Minneapolis is testimony to the way in which co-ordinated private initiatives and public-sector policies invigorated the demand for central area retailing and office space. Tangible evidence of regeneration became visible in numerous prestigious high-rise office buildings, designed by some of the foremost architectural firms in the USA, including Johnson and Burgee, Skidmore Owings & Merrill, Murphy and Jahn, Cesar Pelli and Minoru Yamasaki. Between 1960 and 1980 only four buildings valued at over $20 million had been built, three of which had been developed by the public sector.

In 1978, the value of building permits issued for new construction or refurbishment rose dramatically, by $100 million, representing a rise of over 85 per cent over the previous year. This was to herald a staggering growth in the value of major projects reaching completion in the early 1980s. By 1980, the completion of the 46,000 sq. m. (500,000 sq. ft.) office tower at Washington Square ($35 million) in the Gateway area and the MCDA-assisted Hyatt Regency Hotel ($53 million) marked the start of a major building boom. This resulted in the construction of over 20 office blocks, retail developments, hotels and luxury apartment blocks, each valued at over $20 million.

Between 1983 and 1987, building permits valued at $675 million were issued for central Minneapolis, accounting for nearly a third (31.5 per cent) of the total value for the Twin Cities (Gilhousen, 1988, 7). By far the largest element by value was the office sector, accounting for 53 per cent of all permits. By 1985, downtown Minneapolis was the largest office market in the Metropolitan Area with 1.64 million sq. m. (17.6 million sq. ft.) of space (Gilhousen, 1986). Although the scale of office completions varied annually, reflecting the completion dates of certain large projects, an average of 90,000 sq. m. (1 million sq. ft.) of prime office space was developed annually downtown during the 1980s (Towle Real Estate, 1987). Between 1978 and 1985, the stock of office space in downtown Minneapolis grew by 67 per cent, outpacing development at major rival locations. In comparison, central St Paul's office stock increased by 30 per cent, while the main suburban office concentration on Highway I-494 South expanded by 45 per cent.

This growth in real estate development cannot be attributed solely to local policies. Generous Federal tax code provisions were made in 1981 (Economic Recovery Tax Act) and 1982 providing real-estate accelerated depreciation and tax write-offs for limited partnerships and investment trusts. These established commercial property as the investment choice *par excellence* of the early to mid-1980s and made investment in vacant (non-revenue producing) office space a useful tax shelter for other earnings (Wieffering, 1986; Leitner, 1994). Although this would not specifically have favored downtown commercial development over suburbia, it is probable that the investment market, rather than the user market alone, was a significant factor in determining the scale of development downtown.

Nevertheless, buildings commonly take a number of years to complete and Minneapolis's development boom must have been stimulated in response to factors additional to the 1981 tax code changes because the initiation of development would already have been well in hand. Moreover, user-demand for new office space in central Minneapolis remained very strong during the 1980s, with an annual absorption rate averaging over 65,000 sq. m. (700,000 sq. ft.) between 1984 and 1987. This was reflected in vacancy rates for different classes of building in 1988, at 11.8 per cent for new space, well below that for older buildings, with 'B' class at 13.7 per cent, 'C' class at 20.5 per cent and renovated buildings at 30.6 per cent (Towle Real Estate, 1987).

However, the rejuvenation of central Minneapolis was not entirely successful in countering the trends towards suburban retail growth. While retail sales in downtown Minneapolis had risen from $284 million in 1958 to $390 million in 1982, the real value of sales had actually fallen by 30 per cent (Adams and Van Drasek, 1993). By the late 1980s, the relative decline in vitality of downtown retailing and the prospect of the development of the enormous Mall of America in suburban Bloomington resulted in further plans to improve the attractiveness of Nicollet Mall. Reflecting the intimacy of the relationship between public and private sectors, a senior city planner was seconded to service the Nicollet Mall Implementation Board in order to facilitate the improvement. The concerns also resulted in an MCDA-assisted $56 million retail development on Nicollet Mall to accommodate Saks Fifth Avenue and proposals for an MCDA-backed 83,000 sq. m. (900,000 sq. ft.) private retailing development spanning the Mall. Once again, the protection of real estate values became a major concern, revealed in the submissions by Public Financial Systems to the Nicollet Mall Implementation Board:

> The proposed investment in a New Nicollet Mall would stem the possibility of future decreases in market value due to disinvestment in downtown. Many urban areas are struggling to maintain a healthy economic mix of downtown uses. The commitment to invest in infrastructure improvements will strengthen the retail component downtown and, at a minimum, maintain a healthy level of market values…Both the physical and management improvements for Nicollet Mall and Downtown Minneapolis result in improved salability, or marketability, of the downtown properties and for the conversion of certain properties to higher, better, and more profitable uses.
>
> (Public Financial Systems, 1987, 30)

Nevertheless, the economic and physical rejuvenation of downtown Minneapolis was clearly reflected in the dramatic enhancement of central area assessed land values. The 'Metro 2000' report, jointly published by the City of Minneapolis Planning Department and the Downtown Council, estimated that the actions taken by the public–private partnership had raised the value of downtown property by 11 per cent annually during the previous two decades. Furthermore, based on a comparison with the city block with the highest level of assessed land value, Figure 4.7 shows that between 1973 and 1986 values over a wide zone had experienced considerable increase relative to that peak value. By 1986, the geographical extent of the core (represented by the 25 per cent of peak value iso-line) had widened significantly, as had the 'hard-core' where values exceeded 50 per cent of the assessed value of the peak.

While the assessed land value of the peak block itself more than doubled between 1973 and 1986, rising by 125 per cent, Figure 4.8 reveals that much of the central area

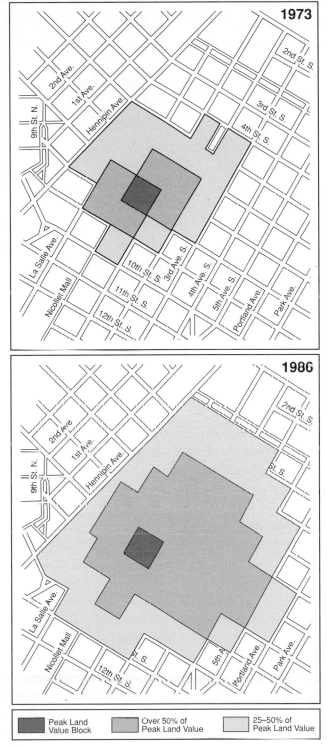

Figure 4.7 *Assessed land market values, 1973 and 1986*

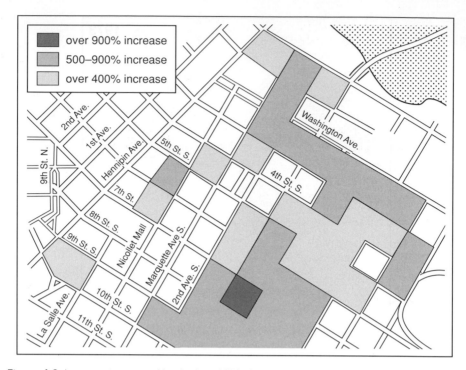

Figure 4.8 *Increases in assessed land values, 1973–86*

experienced substantial increases in assessed land values, exceeding 400 per cent. There were particularly sharp increases in the area near the Government Center, where the public-sector provision of landscaped open space had led to enhanced environmental quality.

4.6 Misgivings and re-appraisal

Between 1980 and 1990, a growing dilemma facing policy makers and planners in the American city was whether to intervene by supporting and subsidizing private development or to permit market forces to dictate outcomes. While a strategy of intervention would enhance the value of private-sector investments, the alternative ran the risk of generating inner-city physical deterioration and a decline in real-estate values. This gave planners little option but to act. Minneapolis city council's concern to protect central-area property values was, perhaps, an inevitable outcome of its financial dependency on downtown real-estate taxes and a perspective in which real-estate value was the key criterion in its evaluation of the success of downtown rejuvenation policy. This was justified on the grounds that maximizing downtown property values permits lower real-estate taxes to be paid by the remaining community (Wieffering, 1986; MCDA, 1987b). They echo Cox's (1981, 445) observations:

> Typically the conflict is between, on the one hand, the urban renewal authority, the downtown booster lobby, and large developers and, on the other hand, residents who will be displaced. The strategy of capital here is to define the community of concern as the central city locked in combat with the other local governments for tax base, and having serious implications for the central city taxpayer.

For the planning authorities, to ensure a high probability of plans' being fulfilled by private-sector agents meant that public-sector goals became tailored to private-sector requirements. The result was the subsidizing of developers and owners of downtown property rights, with public policy aiming to protect and enhance private wealth. Those who lost out directly were the former residents of the downtown periphery, as rising land values drove out low-value occupiers. As early as 1971, the Metropolitan Council (the Twin Cities' regional planning agency) itself had recognized the tendency for downtown developments to drive out lower-grade functions, resulting in a diminishing range of commercial activities (Metropolitan Council, 1971). Historically, the attitude taken by public-sector redevelopment authorities was often unsympathetic towards the remaining residents of the downtown periphery. Between 1973 and the late 1980s, Minneapolis demolished 2,000 of its 5,000 single-room occupancy dwelling units (Leitner, 1990). The residents had been primarily white elderly blue-collar workers who had found low-grade accommodation ranging from rooming houses to 'cage hotels'. Referring to the 1950s Gateway redevelopment, a comprehensive authoritarian planning approach which had indulged planners' desires to regulate, simplify and tidy up the urban landscape and urban land uses (see Knox, 1987), a former senior official on the project commented in a personal interview with one of the authors that:

> It was just prostitutes', bums' and dling-a-ling housing. I'm ashamed to admit it now, but no provision was made for them. We just hoped they'd catch a train to Duluth or somewhere.

4.7 Towards an alternative development strategy for greater community benefit

Throughout the 1980s, Minneapolis had vigorously pursued a policy of public–private partnerships whereby certain developers and corporations obtained significant subsidies to reduce the costs of downtown development. The impact was considerable. Nearly $4 billion in new construction was undertaken in downtown Minneapolis during the 1980s and the downtown tax base rose from 21 per cent of the city's total in 1980 to 42 per cent in 1990. Of the 85 new construction projects which took place in that decade, 40 received public subsidies and 30 of those involved TIF (Fainstein, 1995).

However, despite the success of downtown rejuvenation, visible in the plethora of new commercial buildings, it was becoming increasingly evident by the late 1980s that all was not so well as superficially appeared. A number of economic and social indicators revealed the performance of downtown rejuvenation policy as one of only partial success.

Economically, Minneapolis had continued to be a vibrant central city but, like other major American cities, its downtown core was no longer the centre of the regional economy. As noted earlier, with respect to retailing, although downtown sales had continued to grow, they failed to increase at a pace sufficient to maintain its regional retail dominance. With regard to employment, the number of jobs in downtown Minneapolis had remained fairly stable, with new employers moving into the downtown at roughly the same rate as those who were leaving. For example, 'between 1980 and 1992, the Minneapolis and St Paul central cities showed a one per cent increase in jobs with most of this increase occurring in downtown Minneapolis and concentrated in the services and financial sectors' (Orfield, 1997, 68). However, in terms of the metropolitan area, Minneapolis's share of jobs continued to decline during the 1980s as more new jobs were created in the suburbs.

The national economic recession which commenced in 1988 led to considerable disinvestment in the central city. Downtown Minneapolis lost 15 per cent of its commercial value between 1988 and 1993. Ironically, the state interventions put in place to address the problems of the downtown were identified as one source of the new difficulties resulting from over-development, some analysts claiming that this reduction was due to 'overbuilding' in the 1980s, stimulated by the generous public subsidies (Orfield, 1998).

As problematic as the economic downturn was for the central business area, outside downtown it proved to be even more problematic for the inner city's residential neighbourhoods where jobs declined and commercial vacancy increased. This phenomenon of continuing neighbourhood decline proved the greatest impetus for change in development policy. The indicators of decline in Minneapolis showed that, between 1980 and 1990, employment registered an increase of only 0.7 per cent, median household income rose by just 0.4 per cent, while the number of children under the age of five who were living in poverty rose by 62 per cent (Bureau of Census, 1980, 1990).

The increasing difficulties facing inner-city residents and businesses generated a growing degree of public resentment about the dominance of large-scale downtown property development projects in the city's community development programs. Minneapolis is a city renowned for its active citizen involvement at the neighbourhood level and, by the mid-1980s,

> a number of neighborhood organizations in Minneapolis became active in the electoral arena endorsing five successful candidates for the City Council. Additional Council members with roots in neighborhood organizations were elected in subsequent years. A new grouping thus emerged on the Council in this period that proclaimed its sympathy with neighborhood interests.
>
> (Fainstein, 1995, 14)

Responding to this newly-expressed political strength of the neighbourhoods, a transformed development strategy began to emerge, with city planners and policymakers reassessing traditional policies and considering capturing some of the revenue generated by downtown development and directing it towards the neediest neighbourhoods.

Thereafter, in several ways, urban planning policy shifted towards a more comprehensive strategy which emphasized the redevelopment of both downtown and urban neighbourhoods and the importance of social, economic and design considerations. This reassessment was reflected in the 'Minneapolis Comprehensive Plan' (City of Minneapolis Planning Department, 2000a), a new plan for 'Downtown 2010' (City of Minneapolis Planning Department, 2000b) and in the Minneapolis Neighborhood Revitalization Program (NRP). Taken together, these represented a significant turning-point in the city's urban policy towards a more comprehensive effort to integrate the development needs of downtown with the wider demands of the whole city.

Interestingly, at a time when other cities and nation-states were falling under the spell of neo-liberal philosophies which relied increasingly on public–private partnerships or influencing development through financial incentives and manipulating regulatory frameworks (Healey, 1998), the limitations of such approaches had become increasingly evident in Minneapolis. Notably, the arrangements had tended to confer disproportionately large benefits upon private-sector participants in comparison to those accruing to the wider public.

A number of components of this new thinking will now be outlined.

1.7.1 The Minneapolis Comprehensive Plan

The changing context for development in the 1990s was spelled out in the new *Minneapolis Comprehensive Plan*. It included a downtown goal statement stressing the need to balance new construction and development with other social, economic and design objectives:

> *Goal 7 Market downtown as a place to live, work, play and do business.*
> Downtown, the heart of the economic and cultural capital of the Upper Midwest
> Region, is an exciting and active place that offers the very best qualities and experiences
> that cities can provide. It will continue to be a vibrant place to be, busy with people who
> live, work, shop, dine, and enjoy the special events and unique public attractions. It will
> contain a wide *variety of historical buildings* and the finest contemporary architecture with
> a skyline that continues to serve as a source of civic pride. With parks, plazas and the
> Mississippi River as civic places, it will serve as a model for *other growth centers* in
> Minneapolis which combine significant numbers *jobs, residences, and institutions in high
> amenity areas* well served by transit and other transportation alternatives.
> (City of Minneapolis Planning Department, 2000a, 27)

4.7.2 The Minneapolis Downtown 2010 plan

This planning document provided a 'vision statement' which was the product of a collaborative planning process involving downtown residents, city officials and the Downtown Council. This vision stated that:

> The plan guides development toward a shared vision: a downtown that not only serves as an economic center for the Upper Midwest Region but is also a unique urban community that is constantly alive and filled with people.
>
> (City of Minneapolis Planning Department, 2000b, 2)

Rather than giving emphasis to the single-project approach of previous interventions, the *Downtown 2010 Plan* adopted a broad and comprehensive approach towards addressing social and economic concerns. A key feature would be to use some of the revenues generated by downtown development to address specifically the paucity of affordable housing, the decline in the quality of the housing stock and the need for neighbourhood economic development throughout the city. The means for achieving these goals would be the Minneapolis Neighborhood Revitalization Program (NRP).

4.7.3 The Minneapolis Neighborhood Revitalization Program (NRP)

By the late 1980s, the general perception in the city was that there had been an over-investment in the downtown at the expense of the neighbourhoods and the NRP was intended to reverse this trend. One study estimated that in order to stop the cycle of neighbourhood decline, the city would need to invest $3.2 billion in housing and economic development subsidies over a 20-year period (Fainstein, 1995).

Established in 1990, the NRP set about tackling the legacy of physical decay of housing and declining homeownership which had occurred in the 1970s and 1980s and which accelerated during the recession following 1988. Its ambitious goal was to dedicate $20 million annually for 20 years towards neighbourhood improvements throughout the city. Another NRP objective was to address the disparity in funding for social needs and reallocate resources in favour of the inner-city neighbourhoods. While MCDA expenditures in downtown had risen from $73.8 million in the period 1975–80 to $144 million during 1986–90, spending in the neighbourhoods had fallen from over $495 million to less than $265 million. This marked a reduction from 84 per cent of the total to just 58 per cent (Schwartz and Glickman, 2000).

Originally, funding for the NRP was to be in the ratio of 50 per cent from federal sources, 25 per cent from the state of Minnesota and 25 per cent of local origin. However, it soon became apparent that the federal and state funds were not going to materialize and, having reviewed alternative sources, local officials decided to use TIF revenues generated by the downtown development as a principal source of NRP funds.

As discussed above, TIF funds had already been widely used to support infrastructure improvements (e.g. the development of sewerage, new streets, parking facilities, etc.)

within the downtown and other TIF districts, usually on or near to the development site. However, in the case of the NRP, the strategy would be to transfer 'excess' TIF funds away from the downtown district altogether and direct them towards city-wide NRP projects.

The allocation of NRP funds was the result of a complicated neighbourhood planning process involving the creation of neighbourhood citizen advisory boards, the adoption of neighbourhood plans and the review of these plans by city-wide planning bodies. Within general guidelines, the neighbourhoods were given broad discretion in spending NRP funds on projects such as housing rehabilitation, neighbourhood economic development, social services, parks and recreation facilities. Between 1991 and 1999, housing accounted for 46 per cent of spending, economic developoment some 16 per cent, with parks and recreation accounting for 8 per cent and a further 8 per cent being allocated to human services.

4.7.3.1 The NRP in action: the Stevens Square scheme

The objective of NRP, then, was to redirect a portion of downtown TIF funds to promote revitalization in declining inner residential areas. Stevens Square was an area in which one third of the residents had incomes below the poverty level. With a housing stock dominated by private renting (95 per cent), the area provided affordable housing for some of the city's neediest families and convenient housing for central city workers. It was to become an example of an area that significantly benefited from the use of such TIF-transfer funds which, prior to the NRP and the new downtown policies, would not have been made available.

The Stevens Square Community Organization (SSCO), a community non-profit organization, sought to upgrade the quality of the rental housing while maintaining its affordability by working with the largest landlord in the area, the Stevens Community Associates Limited Partnership (SCA). The SCA owned 23 buildings in the area, comprising 618 apartments, 20 per cent of which were inhabited by low-income, subsidized renters. After intensive planning and negotiating, the SCCO reached agreement with the SCA whereby they would use NRP funds to implement a $14.8 million renovation of these units, with SCA contributing $3.5 million in private funds towards the total costs (Minneapolis Neighborhood Revitalization Program, 2000).

Renovation included the modernization of kitchens, plumbing, roof repairs, etc. in order to bring all units up to housing code standards. In addition, 34 studio and one-bedroom units were converted to 14 two-bedroom units in order to meet the pressing demand for more family accommodation. In return for subsidies, the SCA agreed to maintain rents at levels affordable to low- and moderate-income workers. The project demonstrates how the NRP sought to upgrade deteriorating housing stock and to maintain affordability by creatively utilizing downtown TIF funds, additional local and federal funds together with contributions by developers. Indeed, US Department of Housing and Urban Development Secretary, Andrew Cuomo, praised this project by stating:

> Your success is a testament to what communities can accomplish when they work in partnership with government, private organizations and local agencies and most important, with citizens.
>
> (Minneapolis Neighborhood Revitalization Program, 2000, 8)

4.8 The Target Project: planning reverting to type?

Seemingly, by the late 1990s, the city of Minneapolis had officially adopted a strategic-comprehensive planning approach towards urban revitalization that used downtown development to fund neighbourhood housing, economic development and social policies. Nevertheless, although the 'Downtown 2010 Plan' clearly endorsed this balanced approach towards development, the city council members and the mayor found it difficult to consistently resist pressures from the downtown development community to become personally involved in 'making deals happen'. This had typically involved supporting downtown single-project buildings and providing generous public subsidies to individual developments, regardless of their economic or social merits.

Nowhere was this more evident than in the example of the Target Corporation office building and store project that was initiated for Nicollet Mall in 1996 and finally approved in 2000. In 1996, the Ryan Company, a major downtown developer, persuaded Target Corporation to join with them on a new Nicollet Mall building proposal that would consist of an anchor two-storey (13,935 sq. m. (150,000 sq. ft.)) Target discount store and a 37,160 sq. m. (400,000 sq. ft.) office tower with a store. The exterior of the Target store would be designed to look like a series of century-old store fronts (Meyers, 2001).

However, the public subsidy required to make this single project a reality would be enormous. It demanded over $62 million in public funds to support a scheme with a total development costs of $164 million. In fact, the subsidies relating to land acquisition and underground parking were so great that the projected tax increment from the project were not even sufficient to repay the bonds. Thus, the TIF funds would need to be 'packaged' with funds from two adjacent new office towers, effectively reducing the funds available to the NRP program and other downtown initiatives. As one observer noted:

> The city sold more than $62 million in tax increment financing bonds for the development. The city will need to make bond payments until 2026. If the block had been left to existing uses and private development, property taxes generated by those uses would have gone to the city for development of other uses ... Taxpayers will pay substantially for this project.
>
> (McDonald, 6/10/2001)

Although the rationale for this generous public subsidy was the alleged need to attract a major discount retailer to the downtown, others thought there was also a more overtly political rationale. In 2001, an election was scheduled in which Mayor Sharon Sales Belton and a majority of the council would be running for re-election and many politicians believed that delivering the downtown Target development would be attractive to voters.

The problem with the Target project approach to development was not only that it was questionable economics for public funds to subsidize 38 per cent of the costs of a private building. It also marked a reversion to the previous strategy of making isolated single-project development deals based on 'political' criteria. An alternative development approach had already been established in city council approved plans and policies. But the alleged political advantages of this project overwhelmed the official policy.

However, the Target store controversy did revive the debate about 'who pays and who benefits' when public subsidies are used to support private development. During the first half of 2001 and leading up to the final vote on the council, a lively media debate took place which resulted in the Target project becoming one of the major campaign issues. Ultimately, the Target store project was approved but its much-vaunted political advantages proved illusory as the incumbent mayor was defeated by mayoral candidate R. T. Rybak, an opponent of the Target building, and a majority of the city council was put out of office in the November 2001 local government elections. The voters had strongly endorsed preserving the Neighborhood Revitalization Program and the comprehensive downtown policy. However, if this 'alternative' development strategy is to survive, the new Mayor and Council will need to stand up to formidable political and economic forces favouring a return to the traditional market-driven downtown strategy.

4.9 Conclusion

This chapter illustrates the power of what Greer and Orleans (1962) term the 'parapolitical structure' in shaping public policy. In particular, it lends unequivocal support to Cox's (1973) contention that the downtown business élite constitutes one of the major groups dominating the allocation of public resources in favour of its own area of interest. Cox maintained that the immobility of the investments of the downtown business élite generates a demand for policies aimed at sustaining and increasing the business vitality of the central city. This 'booster lobby' presses for policies which enhance the accessibility of the central city for the downtown workforce and for affluent suburban consumers alike, giving rise to demands for mass transit, freeway construction and the provision of parking facilities. (For a discussion of the city as a growth machine see Molotch, 1976; Logan and Molotch, 1987.) Other policies, such as urban renewal, the development of subsidized convention centres and theatres involve public expenditure in the central city which promote a particular character for downtown, rendering further private-sector investment less risky and more likely.

Ewen's (1978) work in Detroit is also relevant, demonstrating the way leading business families were able to influence policy makers through the civic functions which they undertook. In Minneapolis, the business élite has a high degree of coherence (Galaskiewicz, 1979). It gained considerable influence over civic affairs through the operation of the 'Minneapolis Five Percent Club' founded by leading business families. It comprises an informal philanthropic association of the city's major companies, some 99 firms donating 5 per cent of annual pre-tax profits to civic and community purposes and a further 53 donating 2 per cent (Adams and Van Drasek, 1993). However, in discussing the Hudson company's sponsorship of the Whittier neighborhood revitalization programme Hanson and McNamara (1981, 38) observed that such altruism undoubtedly had an element of self-interest:

> If Minneapolis prospers, the Dayton Hudson stores will prosper too. With their flagship store and their corporate headquarters in downtown Minneapolis, Dayton Hudson puts a high priority on the quality of the central business district and its surrounding neighborhoods ... Thus the Dayton Hudson Corporation has a solid motive for promoting inner-city neighborhood revitalization.

Organizations such as the Chamber of Commerce, the Five Percent Club and the Downtown Council became important vehicles for constructing a dense and cohesive network of business interests in the city and in fostering what Saunders (1979, 316) terms 'institutionalized friendships'. These permit businesses to gain a type of access to those responsible for determining public policy which is qualitatively different from the access available to the public generally. Knox (1987, 293) summarizes the effects of this form of power over public policy:

> Because of the importance of its contribution to the city's economic health in the form of employment and tax revenues, the business community is in an extremely strong bargaining position and, as a result, its interests are often not so much directly expressed as anticipated by politicians and senior bureaucrats, many of whom seek the prestige, legitimacy and patronage which the business élite is able to confer.

This chapter demonstrates that in its power to manipulate the functioning of urban planning to its own advantage, the coalition of private-sector interests represented by the property-owning and business élites in Minneapolis acted in a manner illustrative of the 'manipulated city' hypothesis (Gale and Moore,1975). Acting in conjunction with the political élite and local state administrators, it created a formidable pro-development coalition. This important theme is further developed in the Chapter 9 case-study of Sioux Falls by Breitbach and Mitchell later in this volume.

The chapter also emphasizes the critical support which planning provides for property development and investment (Scott and Roweis, 1977; Healey, 1998). However, in the context of Minneapolis it could hardly be said that planning was 'captured' by private-sector business interests. Rather, effective modern city planning was actually the consequence of private-sector needs and initiatives. It was the creation of a wholehearted sharing by public officials of the same assumptions and long-term goals, as Aschman (1971, 8) noted:

.... the Downtown Council in its earliest days recognized that comprehensive planning in Minneapolis had to be strengthened; and it played a key role in achieving reorganization and major upgrading of status and financing of the Minneapolis planning agency.

Planning in Minneapolis was the creature of private sector needs. With the pre-eminence still given by public policy to the protection of property values, it appears that this is what it has remained. It is encapsulated in a casual observation made during an interview with a senior city planner in Minneapolis, conducted by one of this chapter's authors. Oblivious to the line of homeless people waiting in the cold of a Minnesotan winter's afternoon to gain access to the night shelter off Hennepin Avenue, he explained: 'You see, we don't really have any real problems in Minneapolis. We're planning for a comfortable community here'.

While a review of downtown rejuvenation in Minneapolis supports many of the findings of research into urban political power, these often fail to recognize fully the depth of ideological penetration of the decision-making process in public policy. The nature of power over public policy in the urban arena is more profound than can be revealed by an investigation of either economic strength or political influence, though these two are undoubtedly significant. The question of power runs even more deeply than considerations of the so-called 'second face of power'; the ability to confine and channel conflict and debate towards issues that do not threaten the fundamental bases of the system (Bachrach and Baratz, 1962). It is vital to appreciate the hegemonic role of prevailing ideology. Influence in public affairs is based on a profound belief, common to government officials and public alike, on an unquestioned taken-for-granted faith that capitalist relations of production are not simply the best, but that they represent the *only* possible system. This is a pervasive, unspoken and largely unrecognized power grounded upon an uncritical ideological reflex. It is a conception of 'reality' which directs public policies in certain directions because they are regarded as self-evidently 'realistic', thereby effectively eliminating the potential of alternative radical actions by relegating them to the garbage can of 'unreality' itself.

4.10 Notes

1   The downtown council, which is still consulted periodically regarding proposed planning changes, also provided a vehicle for the voluntary co-ordination of the activities of downtown businesses and real-estate owners. This enables the central area to approach in effectiveness the benefits that accrue from common management of suburban malls.

Acknowledgments

The chapter is partly based on discussions and interviews with senior city planners, senior officials in the Minneapolis Community Development Agency, the Chairman of the Minneapolis Building Owners' and Managers' Association, key commercial property developers and investors, real estate agents and leading private-sector property analysts, together with an analysis of property valuation records from the City Assessor's office. We sincerely thank all those who participated.

5
Planning central Sydney

Pauline McGuirk

5.1 Introduction

Contemporary Sydney is a sprawling, polycentric city which dominates settlement on Australia's eastern seaboard. [1] As the capital of New South Wales (NSW), the most populous state in Australia, Sydney is Australia's biggest city and has long dominated the urban system. Currently, with a population just short of 4 million, the city houses 21 per cent of the nation's population. The city has traditionally had a diversified economy with a strong base in both manufacturing and services, a major port and all the high-order functions and amenities of a regional capital. In the last 35 years, Sydney has progressively taken over from Melbourne as the nation's key site of financial, corporate and political power and has done so in spectacular fashion. The Sydney basin now produces 23 per cent of the nation's GDP and houses more than 275 multinational corporate regional headquarters (O'Neill and McGuirk, 2002). The city has emerged as Australia's gateway city for finance, people, trade and information. Sydney is Australia's global city.

The planning framework has had to contend with the transformation of the role of the city centre from that of a somewhat provincial postcolonial commercial centre to that of the CBD of a global city: a transformation that has generated intense development pressures with which planning instruments have struggled to keep apace. Ashton (1993) famously labelled the city of Sydney (the city's central area) 'the accidental city', its development having been defined more by planning lapses, political manipulation and civic contestation than by systematic planning. The challenges to any systematic approach to planning have since been intensified by Sydney's changing role in the global economy and the related demands to ensure the supply of high-quality real estate, infrastructure, cultural and tourism facilities, and the 'amenity, image and environment' expected of a 'global city' (Hamnett, 2000). These demands underlie the increasingly entrepreneurial nature of the relationship between the planning and development processes, and the increasingly politicized nature of urban planning.

This chapter will trace the direction and achievements of planning and development control in central Sydney in the postwar period, with a particular emphasis on attempts to plan for and regulate commercial office development. This involves examining the often contentious interplay between the planning powers of state government, which provides the strategic planning framework, and Sydney City Council[2], one of more than 40 local governments in metropolitan Sydney which has statutory planning responsibilities for local area planning and development control (see Figure 5.1). The city

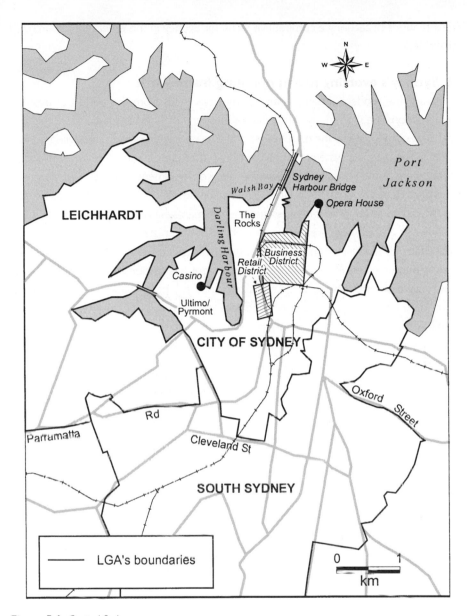

Figure 5.1 *Central Sydney*

of Sydney has always been a strategically important site where the continual struggle between state and local government for political control of planning and land use regulation is evident.[3] It is also a site that has experienced sustained growth, development and intensification of use unparalleled in the metropolitan area. Assessing the uneven successes and failures of planning to control and guide the direction of central city development reveals how planning aspirations have been most successfully achieved insofar as they have coincided with market demands in property development and with state intentions of securing a competitive niche for Sydney in the emergent hierarchy of global cities.

5.2 Sydney's evolving urban planning framework

In Australia, the modern urban planning framework dates from the period of postwar reconstruction. The Commonwealth plan for 'properly planned' postwar economic and social development led to the establishment of a town planning system modelled on that of Britain. The planning system adopted in NSW attempted to use the twin vehicles of metropolitan strategic land-use planning and development control (administered primarily at state government and local government levels respectively) to regulate and co-ordinate the development process and, thus, to manage a rapidly growing metropolis. This two-tier division of power and responsibility between state and local governments has been largely maintained to the present time. Though no metropolitan planning body has ever been created for Sydney, state government has responsibility for any issue or infrastructure of metropolitan range and significance. It is paramount in regulating urban macro-management and strategic metropolitan planning. Local government, by comparison, is a creature of state government, having no formal independent existence or constitutional recognition and highly limited powers and responsibilities. Local planning authorities, of which Sydney City Council (SCC) is one, respond to the policy and strategic planning role of state government (Painter, 1997).

The provision of metropolitan-wide planning for Sydney was initially enabled through the establishment in 1945 of the County of Cumberland Council to prepare a comprehensive metropolitan framework into which local planning schemes were to sit. The Local Government (Town and Country Planning) Amendment Act (1945) gave state government the power to require local councils to prepare local planning schemes in accordance with the metropolitan framework. Once city-wide policies were set, their implementation was left up to local councils according to their planning schemes. These schemes were essentially a collection of restrictive zoning and land sub-division controls according to which development was assessed and development control decisions made. Thus the statutory planning instruments through which strategic aims were to be activated were essentially negative, able to prevent particular uses in particular locations (via zoning) but unable to direct investment by private enterprise. Despite their negatively-oriented planning powers, local government planning schemes and development control mechanisms were intended as the drivers of metropolitan plans. These plans have sought to keep pace with the planning challenges produced by quickening demographic, economic and environmental change and the changing political and ideological framework in which their implementation has been embedded (see Figure 5.2).

Sydney's metropolitan plans have generally struggled to contend with unprecedented population growth and the corresponding pressures of speculative suburban development underpinned by migration, high household formation, the full employment

County of Cumberland Planning Scheme (1948)
Sydney Region Outline Plan (1968)
Sydney into its Third Century: Metropolitan Strategy (1988)
Cities for the Twenty-first Century (1995)
Shaping our Cities (1999)
Shaping Western Sydney (1999)

Figure 5.2 *Metropolitan plans for Sydney 1948–99*

of the postwar 'long boom' and the related demand for home ownership.[4] The plans have tended to be overtaken, if not entirely overwhelmed, by the social, economic and environmental conditions they have had to confront; conditions of unwavering economic expansion (McGuirk and O'Neill, 2002). By the end of the 1970s, however, the climate in which planning operated underwent something of a transformation. The economic downturn took the heat out of the property boom and coincided with growing public demand for participation in decision-making regarding planning and development. This arose from increasing public realization of the political nature of planning and a growing mood of public discontent in Sydney regarding development. More specifically, it arose from growing environmental awareness and public outrage at the extent to which the property boom, particularly in the city centre, had led to the destruction of much of Sydney's heritage landscape, little attention having been paid at either state or local government level to the preservation of historically and socially significant areas.[5]

The 1970s became an era of growing dissent, street protest, innovative social movements and a marked politicization of planning. Together, these contributed to significant reform of the planning legislation which produced the system that currently frames Sydney's planning practice. The Wran Labor state government (1976–86), which engaged seriously with issues of urban planning and built-environment preservation, oversaw a revamping of the legislative framework for environmental planning in the form of the Environmental Planning and Assessment Act (1979). The act aimed to shift the balance away from a development-focused planning system to one capable of considering social and environmental aims and of accommodating a greater level of public participation.

5.2.1 The contemporary legislative framework

The Environmental Planning and Assessment (EP&A) Act (1979), along with the EP&A (Amendment) Act (1997), now frames strategic and statutory planning and development assessment in NSW (see Farrier et al., 1999). State Environmental Planning Policies (SEPPs), Regional Environmental Plans (REPs), Local Environmental Plans (LEPs), and Development Control Plans (DCPs)[6] are the major environmental planning instruments. The plan-making system was devised with the intention that LEPs (produced by local governments) should be consistent with the broader scale of SEPPs and REPS (produced by state governments), all of which should follow the strategic direction of state government's metropolitan strategic planning.[7]

Generally, LEPs determine the development status of a site. They traditionally take the form of a land-use zoning plan for a local government area (LGA), making zoning still the major instrument through which development is regulated.[8] As the authors of LEPs, local councils act as consent authorities for local development. In addition to their development assessment role, under Section 94 of the EP&A Act, LEPs can stipulate contributions from developers for the provision of public amenities and services or the negotiation of other 'material public benefit'. These contributions have been highly unpopular with developers and occasionally have been successfully challenged at the Land and Environment Court. LEPs do, however, have to be signed off by the minister who has significant power to override the plans or to alter them quite substantially if it is believed that matters of state or regional planning significance are involved (Farrier et al., 1999).

REPs and SEPPs are more powerful instruments produced by state government which can be used to shape particular forms of development (e.g. medium-density housing) or to shape development in particular areas (e.g. centrally-located former industrial zones). REPs generally concern planning matters or areas with regional significance, but they are used flexibly. They have been employed to replace LEPs, overruling local councils where state government has wanted either to facilitate a redevelopment or to prevent development in a sensitive area.

SEPPs are reserved for planning matters considered by the minister to be of state significance. They tend to be used incrementally, for instance to amend an LEP to impose special consent requirements, or in a growing precedent, to override local councils. SEPPs have been used to enable the development of specific strategic sites by transferring consent authority from local to state government. For instance, around the the city's central area SEPPs have been used to ensure the development of the Star City Casino and entertainment complex (SEPP 41) and the film and television studio and film-related entertainment complex at Moore Park (SEPP 47). Their use in this guise has, understandably, provoked a good deal of controversy. The Minister can also 'call in' a class of development or a particular development application on the grounds of its state significance, thus resuming planning authority from local government for ministerial determination. Not surprisingly, the balance of power between state and local government – the subject of contentious struggles for many years – is reflected in the balance of power between LEPs, REPs and SEPPs. Recognition of the strategic importance of development in the central city has attracted direct state government intervention into the processes of local plan making by Sydney City Council. All of these mechanisms discussed above, REPs, SEPPs and ministerial call-in powers, have been engaged by state government to supplant local planning control in managing the development of Sydney's CBD and its environs.

Alongside the regulation of the major environmental planning instruments lies heritage legislation. The Heritage Act (1977) provides the main legislative framework for regulating development involving aspects of the built environment with heritage value. Heritage sites are placed on a State Heritage Register.[9] The Heritage Council must be consulted where any listed heritage item is affected either by REPs or LEPs.[10] Many LEPs contain a list of specific heritage items and some identify whole heritage conservation areas. While the minister can direct a listing for a building on the register if it is considered of state heritage significance (s)he also has significant power to overrule these measures. Heritage legislation, though often resisted by developers and, as we will see, sometimes overruled by state government, has been closely involved in inner Sydney's development.

5.2.1.1 Sydney's special case: the central Sydney planning committee

One final layer of planning legislation is pertinent to Sydney's city centre. Central Sydney's status as the symbolic heart of the economy of Australia's most important city and the location of the nation's prime real estate, produces particular planning challenges. These are reflected in the contentious struggles between state and local government over its planning. These struggles have resulted in the formulation of a unique arrangement between state and local government for dealing with central Sydney's development control and strategic planning. The City of Sydney Act (1988) was passed by the Greiner Liberal state government in response to a particularly intense period of conflict over delays in development approvals by Sydney City Council. This act established the Central Sydney Planning Committee (CSPC) made up of the Lord Mayor, two aldermen, the state planning director and four ministerial appointees. The CSPC took over the most significant planning functions of the city council, planning policy, becoming the consent authority for all major development valued at more than A$50m, all non-complying development and any development that the minister requests be dealt with by the CSPC. Hence, all local plan making (LEPs and DCPs) and the assessment of many of the city's development proposals are framed by the recognition of the CBD's unique strategic importance.

5.2.1.2 Provision for public participation

To no small extent, the EP&A Act was a response to civic disquiet about the level of public participation. The Act makes provision for public input into plan-making and, to a lesser extent, the development approval system. There are, however, phases and categories of development from which participation is excluded. In general, the forward planning stages contain the greatest provision.[11] With regard to development assessment, the right to comment on applications for development consent is limited to certain kinds of development. The general position is that unless the minister orders an inquiry into a particular development application, neighbouring owners have no legal right to have their views considered by the local council (Farrier et al., 1999). If a council gives development consent, neighbours have no right of appeal against the decision and there is no general right of 'third-party appeal'. Nor is there right of appeal against the certification of a 'complying' development: development that can proceed once certified by council or by private accredited certifiers as meeting pre-determined development standards. Public participation rights are limited on development deemed to be of state significance, there being no ministerial obligation to notify affected persons. Finally, there is nothing in the NSW legislation which guarantees private property rights or provides for the payment of compensation if the government regulates land use.

Appeals are dealt with by the Land and Environment Court which was created alongside the EP&A Act. Its major role is to determine merit appeals on planning and building decisions by local councils or state government. Appeals against council conditions or refusal of consent can be made with 12 months. Appeal triggers a complete rehearing of the case, with the court taking the council's place. Proceedings have tended to be adversarial and have been criticized heavily by developers for being too costly and lengthy. Significantly, the court is not bound by prevailing council plans and policies such as DCPs, building heights or densities in making its determinations.[12] In practice, the court has allowed major deviations from existing LEPs to be approved and has been used by some developers as a means of circumventing the restrictions of local plans (Searle, 2001).

As we will see, in the increasingly entrepreneurial climate in which planning operates, there has been a growing tendency for controversial developments or simply large developments to be removed from the local government realm and resumed to the control of the state government planning department or directly by the minister. There is no right of appeal, either by the applicant or by objectors, to the Land and Environment Court if the development has been thus called in by the minister. The delays, costs and risks of public participation have made it a contested issue, particularly from the point of view of developers. The Land and Environment Court in particular has played a contentious role. It operates in a highly political environment. State government has several times interfered politically in the judgments of the court. Most recently, state government legislated in 1999 to validate all approvals granted in respect of the development of a heritage-listed wharf at Walsh Bay, immediately west of the CBD, which was under challenge in court proceedings at the time. Critics, notably Sydney city's lord mayor Frank Sartor, have branded the court a 'developers' court', inferring a pro-developer bias in its decisions and arguing for the removal of merit appeals. However, a review of the court's operations in 1999 supported the court, an outcome greeted with great satisfaction within the development community but some disappointment from Frank Sartor and local government planning authorities.

The planning system has grown in its complexity, involving an often bewildering array of state departments, planning agencies, plans, and planning instruments. It has been regarded by the business and development industries as being overly complex. The development lobby, most vociferously represented by the Property Council of Australia, had been exerting strong political pressure for reform of the statutory planning system to enable the fast-tracking of development and a simplified, more flexible planning system, with more uniform and less prescriptive standards, with more private-sector involvement in its administration. The Integrated Development Assessment reforms passed as amendments to the EP&A Act in 1997 went some way towards enacting this deregulatory agenda and providing greater flexibility by streamlining the development assessment system.[13] In 2002, the state planning authority (Planning NSW) is on the verge of introducing further extensive reform of the plan-making process aimed at better articulation between state strategic planning objectives, a strengthened role for regional planning, standardized use of zoning categories and a simplified planning system with greater public consultation in its early stages. The development industry, ever in favour of a simplified system offering more decision-making certainty, has been fulsome in its support of the essence of these proposals.

5.3 At the coalface: planning and city centre development to the 1970s

The central city area has traditionally been paid scant attention in Sydney's metropolitan plans. Nonetheless the city's development trajectory has been shaped by the continuity of decentralization as a core aspiration in successive metropolitan planning strategies. The initial Cumberland County Plan viewed the CBD as unnecessarily congested and aimed to address this through the development of a series of district centres as nodes of local employment, retailing and services. This would serve the twin aims of dispersing the city's industrial and commercial functions while providing employment and services for planned suburban expansion. The role foreseen for the city centre was that of a high-order service centre serving a metropolitan hinterland. Decentralization, in the form of district centres, growth corridors, regional centres and growth nodes, remained central to the broad strategies for shaping metropolitan development. Together, planned decentralization and market-driven suburbanization resulted in the continued dispersal of retail development across the city to such an extent that by 1980 just 10 per cent of Sydney's retail sales took place in the city centre, compared to 50 per cent in 1940. By 1971 only 42 per cent of jobs were found in the CBD, the remainder being dispersed around the city, though concentrated in regional and district centres. The progressive strategic dispersal of these functions created development opportunities for commercial office development in the CBD; a pre-condition for the commercial property boom that would transform inner Sydney in the late 1960s and early 1970s.

The production of Cumberland County plan charged the Sydney City Council with producing a local planning scheme, sympathetic to its major precepts. The scheme produced in 1952 consisted of a conventional zoning approach, separating industrial from residential development and introducing the idea of a floor-space ratio to limit building height. It was heavily criticized by retail and commercial property developers who saw it as restrictive of development potential. Developers tended to see office development as an antidote to the loss of residential population and the dispersal of retailing and industry, so resisted any planning measures that might constrain its potential. As a result of staunch opposition, no formal planning scheme was gazetted until 1971. In the interim, development control (such as it was) was achieved through building regulations and interim development orders, the draft planning scheme operating as merely a land-zoning document.

5.3.1 The development boom 1968–1974

The late 1960s and early 1970s was a period of staggering growth in commercial office development in the city of Sydney which was fuelled by the CBD's transition from a general-purpose city centre to a specialist financial and business services centre. Fed by broader economic prosperity and a massive in-flow of international capital, the property boom of 1968–1974 was the biggest single boom Sydney has experienced (Turnbull, 1999). By its completion, the city had been almost unrecognizably transformed. The physical transformation was served by the lifting, in 1957, of a ban in place since 1912 which prohibited buildings of more than 12 storeys. Lifting the ban allowed the development of the 'big end' of town (Figure 5.2), symbolized initially by the development of the 26-storey AMP tower at Circular Quay (1962) by the Australian Mutual Provident Society, the largest of the life assurance companies. Such developments marked the emergence of Sydney as a lucrative property market to which developers flocked in the following years as property speculation became a rampant city centre phenomenon, fuelled by a combination of financial circumstances.

The availability of finance was assured by the growth of Australia's foreign reserves and the artificially low value of the Australian dollar caused by the fixed exchange system. This produced an in-flow of speculative capital which was enhanced by the government's relaxation of controls on the domestic money supply. Property was one of the obvious sectors into which investment would flow, and was made all the more attractive by the lack of capital gains tax (see Daly, 1982). This attracted a range of speculative property investment, particularly from UK investors, which reached a climax in the early 1970s.

Figure 5.3 Central Sydney – the 'big end' of town

British property developers such as MEPC, Abbey Group and Grosvenor Estate were attracted to Sydney's profitable opportunities. Life assurance companies had their revenues raised by rising inflation and so were attracted to the investment opportunities in the commercial property market, often joining forces with developers in joint ventures.[14] They were followed by pension funds, other financial institutions and even industrial capitalists (see Daly, 1982). Fuelled by Sydney's emergence as a financial hub in the Asia-Pacific region, the demand for office space rose rapidly. Office rents rose from A\$495 per sq. m. (A\$46 per sq. ft.) in 1957 to A\$871 per sq. m. (A\$81 per sq. ft.) in 1971, making commercial property attractive to institutional investors and providing sustained strong incentives to developers to undertake speculative development.

The office stock in the CBD registered major growth. Between 1957 and 1976, 210 new buildings were developed, 84 of them after 1971 (Daly, 1987). Office towers that were to become iconic of Sydney's maturation as an international finance and services centre such as Australia Square (1969), the second AMP tower on the quay (1975) and MLC tower at Martin Place (1977), changed the city's skyline and its position in the

Figure 5.4 *Deep excavation prior to high-rise construction, Sydney CBD*

emergent globalized economic landscape. While development pressures eased between 1974 and the late 1970s with the onset of a global economic slump, the oil crisis, rising inflation and interest rates, this proved to be a temporary lull.

5.3.2 Planning and the boom

This boom period demonstrated the fragility of planning powers and proved to be something of a low point in Sydney's planning history (Ashton, 1993). Sydney City Council (SCC) did little to control the boom in private office development and the Liberal Askin state government (1965–75) was largely pro-development (see Turnbull, 1999). Indeed, the initial phases of the property boom had passed before the city council had even adopted a planning scheme. To address this deficiency, the state government appointed a Planning Commission between 1967 and 1970 to act as consent authority. It is said that the meetings of this commission lasted an average of two to three minutes as commissioners served the unprecedented demand for development approvals with almost unerring consent (Ashton, 1993). In the 22 months of the commission's administration A$300 million of development applications were approved, permitting a huge increase in development activity.

The pro-development emphasis continued when the Civic Reform Association, a coalition of established business interests, was able to win local elections in 1969 and gain control of SCC. Civic Reform was to dominate the local council until the election of a Labor-dominated council in 1980. Planning efforts during the following years were largely focused on dealing with the demands generated by the commercial property boom. In 1971, the Civic Reform-controlled SCC commissioned a strategic plan for the local government area which heightened the office development flurry by encouraging the centralization of office functions and enabling the consolidation of small sites into larger parcels for redevelopment purposes. This was a significant factor in facilitating the development boom and led to the loss of many heritage buildings. For example, Lend Lease's Australia Square (1967) redevelopment site resulted from the amalgamation of 30 properties.

A series of more broadly-focused strategic plans were produced during the 1970s[15] which were structured around action plans for various city precincts but these were very unevenly implemented, few of their objectives being achieved. First, planners were overwhelmed by the sheer momentum of the property boom, so the daily operation of planning functions was consumed by its management. Second, they were hampered by SCC's very limited powers and its reliance on state government co-operation to achieve strategic planning aims.[16] While statutory plans had legal standing, pursuing strategic aims relied on the political support of state government and its various jealously guarded bureaucratic empires. It was initially possible at least to achieve the strategic aim of concentrating the highest density of commercial activity in the central core where developers' interest was most intense.[17] Here, floor–space ratios of 12:1 were allowed and a bonus system operated to enable higher ratios (up to 14:1) through the provision of features such as plazas, links through to other developments, retail space, terraces or escalators. Some commentators suggested that the bonus system, while well exploited by developers seeking greater development intensity, resulted in problematic design outcomes as developments were set back from the street to accommodate under-used retail spaces, building lines were disrupted and fragmented and poorly utilized public spaces proliferated (Ashton, 1993).

In 1974, the strategic plan was updated and attempted to reduce floor–space ratios on the fringes of the CBD. This aim, which was deeply unpopular with the development community, coincided with the dramatic collapse of the commercial property market. The council's attempt to implement this aim was discouraged by state government and, as the property slump continued, the issue of floor–space ratio as a control device was dropped from the subsequent strategic plan (1977). Council planners were largely too disempowered to implement any significant restraints on the scale or the amount of office development. Neville Wran, the pro-development Labor Premier of NSW elected in 1976, was content to see the council remain so and the power balance in planning remained heavily biased in favour of state government.

During this period, the influence of strategic and statutory planning upon the course of city centre development was overwhelmed by the collective influence of property developers. Attempts to disrupt and redirect the momentum of market demand would have been hampered, in any case, by the negatively-oriented nature of local planning powers, the lack of budgetary powers to pursue local planning initiatives and the dependence on state government approval of planning aspirations. More wide-ranging strategic aims were undermined by the politicized nature of city centre planning which became blatantly apparent in the 1960s and 1970s. These decades brought continuing power struggles between the two major parties concerning local government boundaries and attempts to secure political control of the council. Fractious relations between state and local government, particularly where these were of different political complexions, reduced the ability of a relatively weak planning system to achieve strategic aspirations. The council was dominated by the Civic Reform Association between 1969 and 1980 and proved largely sympathetic to the demands of commercial property developers during this time.[18]

The onset of the 1980s brought a new context for the city's development. Significant transformation in the city's economic base saw Sydney attain an ascendant position in global circuits of capital, as a fledgling global city. Economic transformation was paralleled by the emergence of a new political climate of neo-liberalism. Economic and political change provided a new climate for planning, in which planning's role and, some would argue, its effectiveness have been significantly modified.

5.4 New roles and new planning regimes: planning central Sydney since the 1980s
Economic globalization and industry rationalization have transformed Sydney's fortunes since the 1980s and this has coincided with the emergence and eventual prominence of

a political climate of neo-liberalism. Successive federal governments, activated by a neo-liberal inspired micro-economic reform agenda, have imposed public spending restraints on the states and driven the new philosophy of 'smaller, more efficient government' creating a backdrop for the practice of urban planning of fiscal austerity and public-sector corporatization (Bell, 1997).

The new political and economic conditions created changing strategic directions in urban planning and its desired spatial and social outcomes. They brought a re-orientation of state intervention away from a broadly social democratic tradition towards a new neo-liberal tradition favouring supply-side, facilitative interventions which provide market support above demand-side regulatory interventions (see Gleeson and Low, 2000). For urban planning, this has meant a shift away from an emphasis on achieving equitable urban outcomes, the vision of the equitable city; towards facilitating efficient urban outcomes, the vision of the productive city (see Peel, 1995). The tenor of public policy debates has paralleled this shift, increasingly linking urban policy and urban change to processes of economic restructuring and globalization.

These changed politico-economic conditions have also impinged significantly on the mechanisms through which urban planning aims are pursued. As well as attempting to achieve social and environmental aims, land-use planning has been more closely aligned

Figure 5.5 *Woolloomooloo, Sydney. Sandwiched between the central business district and the commercial area of King's Cross and 'ripe' for upgrading, large-scale commercial redevelopment was prevented by opposition from residents and a 'Green Ban' by the Builders' Laborers' Federation*

with economic planning, providing the strategic settings and regulatory framework for private economic growth. More specifically, it has increasingly been harnessed to the effort to secure Sydney's position in the emergent global space economy. Planning has been enlisted, to some degree, as a form of product enhancement, maximizing amenity, environment and image in order to project Sydney's global city aspirations. This shift has brought increasingly entrepreneurial planning practices and techniques to the fore, along with new pressures for an integrated and flexible approach to strategic and statutory planning. This has meant (i) a heightened concern for managing the costs of urban growth (as a factor in economic productivity), (ii) an increased emphasis on attracting mobile capital investment, via the creation of required supply-side conditions through increasingly entrepreneurial means and, as outlined above, (iii) increased pressure to create simplified, flexible and increasingly privatized forms of planning regulation and development control. These shifts have produced a more complex setting in which planning operates.

In this context, relations between state and local governments have become more controversial as local governments have been given more substantial roles, particularly with regard to local economic development,[19] as entrepreneurial planning has produced new forms of partnership between government and the private sector and, more contentiously, as state government has been prepared to override local concerns in order to pursue broader planning outcomes. These shifts have all taken place against a background of intensified public interest in the politics of development and environment regulation (Freestone, 2000).

5.4.1 City centre development

Despite the impact of cyclical fluctuations, the period since the 1980s has essentially been one of sustained economic growth and demand for property in central Sydney. The office market had collapsed in 1974 leaving 18 per cent of the CBD office space vacant by 1976. Vacancy rates dropped quickly with market recovery and stayed low, at around 3–4 per cent, until 1989. These boom times coincided with the city's emerging role as a key financial centre in the Asia Pacific as a highly internationalized service-centred economy (Daly, 1987).

The early 1980s brought a major slowdown in economic expansion as Sydney's metropolitan economy underwent significant restructuring. Between 1970 and 1985, Sydney lost 178,000 manufacturing-sector jobs as many industrial plants in the central industrial area closed and others rationalized as they relocated to the suburbs. Between 1981 and 1991, manufacturing employment declined a further 25 per cent, from 274,849 to 208,823 jobs (Fagan, 2000). By contrast, Sydney benefitted from globalization and its role as a national gateway, attracting regional headquarters and expansive information-based services and industries. In particular, the information service sector (finance, property and business services, communications, education and administration) witnessed a remarkable expansion. Employment in finance, property and business services grew from 9.5 per cent to 15.4 per cent between 1981 and 1991, as the city's growing role as a node in international circuits of accumulation took effect.[20] These activities have been particularly concentrated in the CBD and, despite the influence of boom/slump cycles, have continued to drive demand for intensive office development.

The development of central Sydney's built environment has been dramatically re-shaped by the demands produced by global economic integration. Sydney, most particularly the CBD, has emerged as an international gateway city for finance, people, trade and information connecting both the local and national economies to global accumulation flows. This role has created considerable demand for office and residential space, for hotel development and for the forms of 'spectacular' consumption sites often demanded of cities with a global role. The accompanying intense development pressures have created challenges which have occasionally overwhelmed existing planning instruments and seen them replaced by new forms of entrepreneurial and privatized planning.

The early 1980s brought East Asian investment interest in property, initially concentrated in the hotel and retail sectors particularly. In 1983, the state government commissioned a report on the prospects of Sydney's becoming a Pacific financial centre; something which drove office development and investment. Notably, the city began to surpass Melbourne's status both as a financial centre and a location for corporate head office activity. By 1990, Sydney accounted for four-fifths of foreign bank head offices and had a strong presence of non-banking sector financial institutions. Likewise, Sydney had also attracted 60 of the top 100 Australian corporate headquarters. A key role as a centre for the regional headquarters of multinational corporations would soon follow.

The process of expansion had been driven by financial deregulation[21] which created a property boom in downtown Sydney, increasingly the hub through which the global flows of information and finance connecting the Australian and global economies are channelled.[22] The growing ascendency of finance capital brought a sustained inflow of foreign capital into commercial property investment through the 1990s. This market for office property was stoked by the decision in 1993 of the state government to launch a programme to attract Asia-Pacific regional head offices of international service corporations to Sydney.

Continuing growth in the services sector also sustained the demand for office space. Service sector employment rose in the 1980s in response to the demands of businesses for specialized financial and legal services. Between 1991/2 and 1996/7 the workforce in the City of Sydney increased by 30 000 (15 per cent), 85.3 per cent of whom were employed in office activities. By 2000, 54.6 per cent of the City of Sydney's workforce was employed in financial and business services which are particularly attracted to the 'big end' of town and its premium-grade office space. Major developments of premium office towers built in this period include those listed in Table 5.1. By 1997, offices comprised 41.3 per cent of central city floor space (SCC ,1997) and the total office stock reached 4.5 million sq. m. (48 million sq. ft.) by 2002 with another 800,000 sq. m. (8.6 million sq. ft.) located at North Sydney (SCC, 2001).[23]

Table 5.1 *Premium office towers, Sydney CBD, 1985–2000*

Building	Date of completion	Floor Area (Sq m)
State Bank Centre	1985	58 630
Grosvenor Place	1987	90 000
IBM Centre 1	1991	45 000
No 1 O'Connell	1991	35 000
Chifley Tower	1992	90 000
Governor Phillip Tower	1993	55 000
DCM Tower	1998	27 000
Darling Park	1999	59 600
400 George St	1999	58 700
Aurora Place	2000	49 000
Angel Place	2000	55 000
Citigroup Centre	2000	101 300

This new generation of buildings now dominates the city skyline and reflects the globalization of Sydney's commercial property development and investment process with its intimate ties to internationalized financial markets. This brought a wide variety of established development companies, institutional investor-developers and developer consortia of national and international origin into play in Sydney's property markets. Figure 5.6 illustrates the dominance of financial interests in the ownership of prime CBD office space by 2001. The insertion of the CBD into global circuits of property investment has sustained demand and created pressure from property interests for a pro-development stance in local planning and development control activities.

Office development opportunities have, of course, waned cyclically in the Sydney CBD but have been replaced by development opportunities in the residential sector. The economic recession of the late 1980s and early 1990s meant that there was an oversupply of commercial floorspace in the city and a cyclical slump in office development activity ensued (see Figure 5.7). Between 1989 and 1992, a major slump resulted in rising vacancy rates to levels not seen since the mid 1970s, the office market remaining sluggish until 1997. During this period, development opportunities were found in high-rise residential units. Several vacant office buildings were converted and many new apartment buildings were constructed, tapping into an unsuspected demand and starting something of a flood of population to the inner city.[24]

In 1992, just 2.5 per cent of Sydney's dwelling commencements had been located in the inner city. By 1996, this had risen to 17.5 per cent with sustained high levels of activity continuing subsequently. Residential building approvals in Sydney and neighbouring South Sydney increased from fewer than 500 in 1989–90 to approximately 4,000 in 1999–2000. Between 1994 and 1999, 65 residential development schemes were constructed in the city, predominantly in converted or newly constructed high-rise towers such as the 44-storey Horizon Tower in Darlinghurst, the 50-storey Century

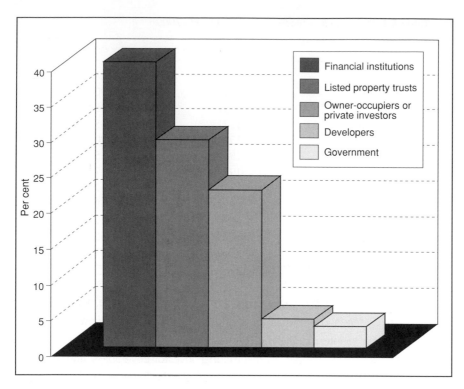

Figure 5.6 *Ownership of CBD prime office space, 2001. (Source: 'Axiss Australia' : The Prime Office Market in Australia, 2001)*

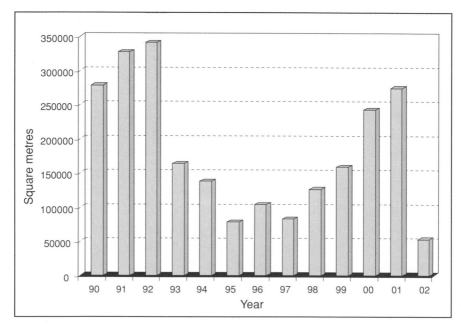

Figure 5.7 *Additions to office space, Sydney CDB, 1990–2002. (Source: Property Council of Australia)*

Tower on Pitt St, the converted 36-storey Observatory Towers at Circular Quay, or The Peak, a 46-storey tower in Chinatown. Many more residential developments took place on obsolete industrial sites freed up by the loss of manufacturing from the inner industrial areas: the CSR plant in Ultimo-Pyrmont, the Unilever site at Balmain, the glass bottle manufacturing plant on South Dowling Street, the Wharves at Woolloomolloo and Walsh Bay. Between 1996 and 2001, 8,707 dwellings were completed in the City of Sydney (SCC, 2002).

Finally, this period also witnessed a major expansion in the development of international hotels[25] and tourist-oriented retailing in the CBD as Sydney gained a more established position on international tourism circuits and became host to the world hallmark event of the Olympic Games in 2000. Between 1989 and 1992, hotel investment in Sydney amounted to A$1.588 billion, with a further A$746m being invested between 1993 and 1998, almost all of which was concentrated in the central area (Daly, 1998).

Hotel development was matched by a changing retail role for Sydney from that of a regional centre to one dominated by specialist functions, with a particular focus on tourist-oriented 'festival' shopping.[26] For instance, the Malaysian development corporation Ipoh redeveloped the heritage Queen Victoria Building in 1986, as a specialist shopping mall. Its redeveloped Strand Arcade, off Pitt Street, served a similar role. The convention/entertainments/retail complex at Darling Harbour was developed by the state government in the late 1980s. The Pitt Street Mall, pedestrianized in 1987, has remained the main functional shopping core it but has been joined by an agglomeration

Figure 5.8 Darling Harbour: a convention, entertainments and retail complex developed in the late 1980s by the state government as a 'bicentennial gift to the nation'

of high fashion designer stores clustering on parallel Castlereagh Street,[27] where target consumers are likely to be international tourists as well as locals. Significant retail elements have also been built into the lower floors of mixed-use office developments, most notably seven levels of retail space beneath 12 commercial levels at the Skygarden on Castlereagh Street and two floors of exclusive retail space at the 55-storey Chifley Tower.

In total, between 1994 and 1999, A\$3 billion of development was completed in the City of Sydney, in the form of 15 hotels, 12 serviced-apartment developments, 35 commercial developments (office, retail and entertainment) and 65 residential developments (7,653 units) (SCC, 2000). As Connell (2000: 325) has suggested, 'the rise of finance, in association with global economic growth, is continuing to restructure the city ... in a seemingly endless process of redevelopment'; redevelopment that has brought ebbs and flows of pressure on planning authorities to attract and facilitate development, to accommodate the new space demands being generated by Sydney's evolving role in the global economy and to manage the conflicting demands of developers, public opinion and political interference in Australia's most strategically important development market.

5.4.1.1 City centre planning
The controversial nature of city centre planning has endured, as has the struggle for political control over this iconic location and the development fortunes of its premium real estate. Strategic plans had been drawn up regularly for the central area since the early 1970s. However, by the 1980s these had had little effect and city planning was essentially 'in chaos' (Ashton, 1993) due to the limited powers held by Sydney City Council, the range of other authorities with some planning powers and the tendency for political interference from the state government if city council policies or processes were thought to restrict development interests.

This was infamously the case in the late 1980s. Recession was evident, economic restructuring was taking a toll and planning of Australia's Bicenntenial celebrations in 1988 was underway. The Labor state government was keen to allow maximum development in the city. SCC, at this stage, had a number of Independent councillors who opposed this unchecked enthusiasm for development and had delayed a series of proposals for hotel developments, drawing harsh criticism from various ministers. Several significant heritage losses in the 1980s had hardened their resolve.[28] When the councillors would not back down, Labor Party Premier Unsworth sacked the city council arguing that 'the city of Sydney perhaps warrants a different form of government ... we believe [the sacking] was warranted in the interests of Sydney, in the interests of New South Wales, and in the interests of Australia' (*Sydney Morning Herald*, 27. 2. 1987). A three-person commission was appointed to oversee the city's development assessment over the following two years, a political move aimed at injecting wider-scaled strategic considerations in to the city's development control (Searle, 1998).

Not surprisingly, a series of high-rise office towers were approved in this period and developed in the early 1990s prior to the property slump. The commission-style control of planning was given more permanent status by the City of Sydney Act (1988), administered by the new Greiner Liberal state government. This established the Central Sydney Planning Committee (CSPC) which took over the most significant planning functions of the city council; planning policy and development assessment on all major developments. Ashton (1993, 116) characterized SCC after this point as 'a civic watch-dog on a very short leash', having the least powerful planning processes of any LGA in Sydney. By the late 1980s, the city faced a phase of property slump in which any perceived obstacles to proposed development were frowned upon, at least by developers and state government.

Following the planning committee's appointment, it devised the 1988 Central Sydney Strategy with an accompanying LEP and DCP. These plans were generally based around aspirations for urban quality which were 'loosely framed' (Ashton 1993, 118). The details of how such qualities should be achieved through development standards were vague and judgements on planning approvals were largely to be made 'according to merit'. Clearly, these were plans devised to be flexible and to have the capacity to deliver planning approvals for significant development proposals.

While the DCP was conservative in many respects, it included a floor–space ratio of 15:1, higher than that allowable in Manhattan. Furthermore, the accompanying heritage plan was widely criticized because of the ease with which it could be modified. Items could be (and were) removed by the Minister for Planning without recourse to legal measures. Indeed, 300 buildings were excised from the Central Sydney Heritage Inventory by the CSPC in the early 1990s as a result of pressure from development interests (Searle, 1998).

Planners, councillors and public interest groups were highly critical of the concentration of power represented by the CSPC in the first instance, but also of vagueness of the CSPC's plans, their scope for manipulation in support of development and the difficulty of holding 'merit judgements' up to public accountability. Criticism resulted in the calling of a Commission of Inquiry into the planning instruments drawn up by the CSPC. In 1992, the Simpson Commission brought down a highly critical assessment which argued that the Committee's LEP and DCP offered an inadequate basis for future planning and development control and recommended their abandonment (see Simpson, 1994).

By this stage, Independent councillor Frank Sartor had been elected mayor, along with strong Independent representation amongst SCC councillors.[29] This opened the way for the city's LEP and DCP to be re-drafted according to a new agenda. The Independents had come to power on a 'living city' election agenda, aiming to produce a city centre that, while targeting commercial growth, tourism and hotel development and sustaining CBD retailing, was also concerned with residential revitalization and the conservation of heritage (Turnbull, 1999). This agenda sought to bring vibrancy back to the inner city; a quality thought to have been badly damaged by the developments of the 1970s which had left windswept streetscapes, deserted after the close of business. By also focusing on cultural activities, a diversity of uses and the quality of city public spaces, it aimed to bring people back to the city.[30]

Thus, since Sartor's election, pro-development pressure on planning has been tempered by strong amenity and heritage groups at local council level, which has mitigated the sustained pro-development approach of the 1970s and 1980s (Turnbull, 1999). The living city agenda could be translated through development control decisions insofar as SCC still had consent authority. Sartor came to power as the worst of the commercial property slump passed and as interest in residential development in inner Sydney rose. Sartor's council worked to redraft the LEP and DCP (approved in 1996) to include more attention to city form, urban design, to accommodate residential development and to include a more sympathetic heritage plan which allowed increased floor space ratios for developments if heritage floor space was protected. Planning controls also capped maximum building height at 228 m. (the current height of Governor Phillip Tower). In this instance, then, the alignment of market and political circumstances and community support enabled significant local planning achievement.

Indeed, in 1997, the CSPC's powers were restructured somewhat to increase council representation and return some plan-making powers to the council with respect to heritage planning (Searle, 1998). This allowed the council to add 43 heritage buildings to the protected list. Since that time, plans have operated in an increasingly negotiative manner, representing something of the growing flexibility in the approach to achieving planned outcomes. For example, some major office developments have been permitted on the condition that they include public facilities as planning gain. Thus, AMP's development at Angel Place included a concert auditorium. The Telstra office complex at 400 George Street was encouraged to include a shopping complex that created essentially a new street, leading shoppers through the massive complex from George Street into the main shopping mall at Pitt Street.

5.5 Evaluating central Sydney planning

Any evaluation of the planning achievements in central Sydney during the 1980s and 1990s must be underpinned by a consideration of the heightened tension between state and local governments and the distribution of their respective planning powers. The increasingly neo-liberal flavour of the political climate in which planning practice is embedded has already been noted. This climate progressively infused the public sector, at state and local levels, with an entrepreneurial approach to governance in all its forms. With regard to planning activity and central Sydney, these impulses have been heightened by the desire to shape and promote Sydney as a global city and to insert the city successfully into the circuit of global place competition to attract and hold mobile investment. These entrepreneurial tasks of planning for 'global Sydney' have been adopted across state- and local-government planning authorities. However, their adoption at the local level has been tempered by local government's greater direct accountability to its local constituency; a constituency that has often been opposed to the forms of development pursued by state government's entrepreneurial planning initiatives. State government's response has been one of selectively centralizing planning power and using facilitating legislation to create attractive investment and development environments; something that local government and its constituents have tended to resent. The impact on planning's powers to shape the course of development has been to render it increasingly susceptible to the logic of the market (Gleeson and Low, 2000).

The success of SCC's living city agenda was undoubtedly assisted by market forces within the property development sector, coinciding with a slump in office development. Developers had been exerting considerable pressure on local councils to rezone industrial land in the central area for commercial and residential uses. When demand for office space collapsed and residential development offered a profitable alternative, a residential development boom commenced in inner Sydney. This was further buoyed by state government planning support of medium- to high-density large-scale residential developments in locations such as Ultimo-Pyrmont, via the strategic use of REPs and SEPPs.

Over the course of the 1990s, Ultimo-Pyrmont, immediately west of the CBD, has been redeveloped at high residential densities and has attracted the interest of the major national residential property developers, Meriton, Mirvac and Lend Lease, all of whom have undertaken very large-scale redevelopments on the peninsula.[31] Demand for these developments and for the high-tech and Internet industries that have also been established nearby has been fuelled by Sydney's economic transformation towards global city status. Demand has emerged, in particular, from a growing pool of élite CBD service workers who constitute a Generation-X market niche for inner-city residential development. Such developments have produced a new residential population in the inner city who have been generally supportive of further rezoning for residential development.

Inner-city residential development, successfully promoted by the living city initiative, has been directly challenged as a result of changing market conditions. As office vacancy rates dipped below 7.5 per cent in 1997 marking a recovery in the office property market, developers became increasingly concerned, first, over the 'sterilization' of economically productive land by its residential zoning and, second, over the problems that can arise when residential development imports a voting population with demands that often conflict with those of commercial property development. The Property Council of Australia, for instance, has been vociferous in its critique of the conversion of development sites around the central area from the most profitable commercial and high-tech industrial uses to the less profitable residential functions. The 3:1 floor space bonus for residential development offered by the 1996 LEP was particularly opposed by commercial development interests, as were the lower floor space allowances permitted in Chinatown: a move designed to prevent new high-rise development from overwhelming the character of the neighbourhood (Searle, 1998). Nonetheless, both sets of controls have been maintained to date.

Although the 1996 LEP remains the legal framework for development control in the city, real power for approving major projects still lies with the CSPC. Beyond this, there has been a very clear centralization of control over projects deemed to be of state or metropolitan significance. This centralization demonstrates the entrepreneurial direction of planning control as the state government has regularly acted to reign in local government's role in planning, usually where local concerns were considered to threaten likely development investment. Various state governments since the Wran/Unsworth Labor government (1976–88) have shifted planning authority over specified geographical zones to special purpose development corporations, deployed state environmental planning policies to deal with specific development areas and used special legislation to override the EP&A Act. Most notably, these powers have been called upon to support and facilitate large-scale redevelopment projects, particularly those concerned with consumption and entertainment/tourism facilities and infrastructural projects with the capacity to act as show case developments for an emerging global city (Searle and Bounds, 1999). This centralization of power has had a profound impact on the regulation of the built environment of the central area.

State governments have desired and actively sought a 'global city' profile for Sydney. This desire has translated into incentives to development and the deployment of state capital and resources, primarily public land, to attract development and investment capital. These all implicate planning in place-marketing and city-imaging activities, as Sydney has boosted its claims to global city status and become actively embroiled in place competition. In a variety of examples of such pro-development support, and in addition to the powers of the CSPC in the CBD, state governments have overridden councils in the central city, underwritten private-sector development risk with state subsidies, supplied public lands and used various means to fast-track development through the approval process. This pro-development action has often sat in an uncomfortable tension with the more measured approach of SCC to city-centre development.

By 1994, Webber (1994, 217) could argue that 'in Sydney, the planning and development control powers of the city council have been so extensively compromised by the measures imposed incrementally by successive State Governments, that it can no longer exercise cohesive planning powers'. The first major incidence of rescinding local consent authority for a significant area and consigning it to the minister occurred in 1968 with the formation of the Sydney Cove Redevelopment Authority (SCRA) to oversee the redevelopment of the Rocks.[32] Then, in 1984, the Wran government aimed to redevelop underutilized railway goods yards at Darling Harbour (see Figure 5.8) as a convention/entertainment/retail centre as a 'bicenntenial gift to the nation'. Fearful of local council opposition and to circumvent planning regulation, a special Act of Parliament was passed to create the Darling Harbour Authority (1984) which could operate outside the confines of council planning controls. Following massive public opposition, when the council threatened to refuse permission to a privately developed monorail linking the development to the CBD, the state government again legislated to exempt the monorail from the Planning Act and remove council powers of refusal.

Another development corporation, the City West Development Corporation (CWDC), was established in 1992. The CWDC drew on generous funding from the federal government's Building Better Cities Scheme.[33] It aimed to drive the state-sponsored redevelopment of the former industrial inner suburb of Ultimo-Pyrmont. Here, redevelopment was successfully achieved through the suspension of local council planning authority over large areas of the peninsula. Redevelopment was fast-tracked, first, by an REP (No. 26) which established planning principles and controls for the area as a whole, outside the LEP-process of local council. The REP aspired to tripling the existing population and generating 50 000 jobs over a 20–30 year period (Hillier and Searle, 1995). It then selected key development sites for which master plans were prepared and in which development proposals required approval of the minister alone. These were predominantly high-value waterfront sites and were often government owned and thus subject to significant pressure to generate revenue for state government coffers.

In the same neighbourhood, SEPP 41 was then used in 1994 to permit the development of a large-scale casino on the former Pyrmont power station site, against heavy opposition from many in the local community. The state government considered the casino, which operates for 24 hours a day on every day of the year, to complement the entertainment and tourism complex in neighbouring Darling Harbour and to constitute a requisite element in the armoury of any city claiming global status.

In the cases of both Darling Harbour and Ultimo-Pyrmont, the prevailing climate of fiscal constraint heightened the pressure to redevelop state-owned land in order to generate revenue. The SEPP was even allowed to prevail over REP 26, should its development contravene that plan's aim to encourage quality residential development.[34]

In 1998, the largest special purpose authority ever created in Sydney was formed: the Sydney Harbour Foreshore Authority (SHFA). It absorbed the SCRA and the CWDC into a state-government agency with overall co-ordinating responsibility for the planning and development of Sydney Harbour. The SHFA is supported by SEPP 56, which empowers the authority to co-ordinate planning and development of land on the foreshore and establishes guiding principles for development. Local planning mechanisms have been overruled by master plans for key foreshore sites. These have to be approved by state government before they can be adopted by local councils. Fourteen individual sites along the harbour have been listed as of state significance and consent authority over them has been resumed to the minister. A further 24 sites are listed as of strategic significance. All 38 require master plans to be drawn up to oversee their management; a move which brings development control on these sites into the ministerial realm. In the case of the SHFA, the agenda is more complex than the pro-development emphases of other authorities. It is obliged to contend with environmental, conservation and urban design issues in, arguably, the nation's most symbolic and high profile site.

Special purpose authorities have not been the only means through which state governments have assisted property development. Planning for the redevelopment of Walsh Bay, to the west of the Harbour Bridge and the Rocks, demonstrated a willingness to override both planning and heritage protection authorities in order to enable development. The Walsh Bay Wharves, which were owned by the government's Maritime Services Board, were placed under a Permanent Conservation Order by the NSW Heritage Council in 1987 after several years of proposals to develop a mixed used tourism, retail, office and residential complex appeared to threaten their survival. In 1989, the state government devised an REP (No. 16) to enable adaptive re-use (involving a change of use with limited building alterations) and removed all council control by appointing the Director-General of the Department of Urban Affairs and Planning as the consent authority.

Over the following years, private-sector development capital was sought which would undertake redevelopment and, in 1996, a proposal was finally agreed.[35] The Walsh Bay Redevelopment Masterplan was approved in 1998, having been delayed because approval from the Heritage Council had to be secured for the demolition of some parts of the wharves. When the National Trust appealed against these approvals at the Land and Environment Court, the state government, determined to push ahead with the development, introduced special legislation. The Walsh Bay Development (Special Provisions) Act (1999) validated all approvals granted in respect of the development, terminated the National Trust's court action, designated the Minister as the appropriate consent authority and prevented appeals against any future ministerial decision at Walsh Bay. The largest heritage waterfront redevelopment in the southern hemisphere was thus able to proceed.

Figure 5.9 *Redevelopment with façade retention, Castlereagh Street*

5.6 Conclusion

Sydney's planning is driven by a complex array of planning instruments and agencies that have aimed to control and guide the direction of property development in line with varied spatial and social aims. These have been shifting aims, whose transformation has paralleled the changing political context and the dramatic shift in the city's position in the global space economy.

By the turn of the century, central Sydney's planning involved a plethora of agencies and authorities. These include the SHFA, the SCC, the CSPC, the Heritage Council, the Land and Environment Court, Planning NSW and the Minister for Planning. The tensions over the distribution of planning authority between these agencies and across tiers of government suggest that planning's successes and failures are as much determined by power struggles within the state as they are by struggles between state and market. Sydney's unique position in the Australian economy and its crucial role as gateway to the global economy have heightened the stakes in these struggles, giving a fragility to planning's attempts to shape development in directions other than those suggested by the market.

Local government has had some significant successes, such as Sydney City Council's living city agenda. Nonetheless, it is still caught in the legislative trap of negative planning control which limits its capacities to achieve desired planning outcomes and restricts it to a reactive role. By contrast, state governments have the capacity and legislative authority to make the sorts of proactive market interventions that can effectively shape development outcomes. To date, these interventions have primarily had a market-supporting role, entangled as they have been in the project of Sydney's economic development and global integration. As their role in the development of central Sydney demonstrates, they have involved an increased privatization of planning, new forms of facilitative partnerships with private-sector development interests and a move to more flexible approaches to planning regulation.

These approaches have undoubtedly empowered planning mechanisms to achieve the desired aims. Indeed, the active involvement of state planning authorities in the dramatic physical transformation of vast tracts of inner Sydney in the 1990s is testament to this. Crucially though, this has been achieved at the cost of suspending conventional planning instruments and, hence, suspending the democratic element of local planning and its participatory mechanisms. These tactics have produced growing public conflict and angst concerning the quality of the urban environment and the limited degree to which public participation is possible in decision-making which will determine the future course of Sydney's development. Not everyone is impressed by the physical and social landscapes that result from the trend of developing the city as a series of isolated masterplanned fragments.

The challenge that faces planning, then, is twofold. The first is how to harness the empowering capacity of proactive intervention and new flexible relations with the private sector to more social democratic goals that will often exceed the narrow confines of market criteria. The second is how to achieve this in a democratic and inclusive manner that enables local government, the most accountable and accessible tier of government, to develop an enhanced role in planning its locality's development. Debate about Sydney's planning future is likely to revolve around these issues for some time.

5.7 Notes

1. It is the world's fifty-sixth most populous city but covers the eleventh largest surface area (Connell, 2000).

2. Sydney City Council has jurisdiction over the central area of Sydney, broadly contiguous with the CBD. Sydney still has no metropolitan planning authority.

3. The city of Sydney has experienced numerous changes to its electoral boundaries over the years in multiple attempts, orchestrated both by Liberal and Labor state governments, to secure its political control by a council dominated by one or other of the major parties. The most recent inquiry into the relocation of local government electoral boundaries took place in 2002 and has resulted in the adoption of a 'no-change' option, following a court challenge from South Sydney Council which opposed proposed council mergers.

4. For example, between 1947 and 1971 population expanded by twice the number anticipated by the Cumberland County Plan.

5. Most significant here was the radical environmental politics practised by the Builders' Labourers Federation. Through the 1970s, this militant union joined forces with various residents' and environmental groups to place 'Green Bans' on controversial development proposals, banning their members' participation in these developments' construction. The BLF's most renowned victory came over the Askin Government's proposals to redevelop Sydney's historic Rocks area as a complex of high-rise office towers and hotels. This government had formed the Sydney Cove Redevelopment Authority in 1968 to oversee the redevelopment and removed it from local council jurisdiction. A BLF Green Ban in conjunction with local resident activism defeated the proposal. In this instance, its consciousness-raising impact was arguably just as influential and important as its saving of the Rocks.

6. DCPs deal with similar matters to LEPs but at a more detailed level. However, unlike SEPPs, REPs and LEPs, they are advisory documents and are not legally binding. They simply outline matters that have to be taken into consideration before a development can be approved.

7. In reality there have been many problems in streamlining the strategic intentions of LEPs with those of REPs and SEPPs and there exists a fractious history of conflicting, or as least non-complimentary, intentions and decisions at these various scales.

8. For each zone, a distinction is drawn in how a development should be assessed between (i) exempt development: minor development that may be carried out without development consent; (ii) complying development: development that can proceed once certified by council or by private accredited certifiers as meeting predetermined development standards; (iii) development which requires a consent, in which case a development application must be made and (iv) state-significant development for which the minister is the consent authority. The category of prohibited development which existed until the 1997 amendment to the EP&A Act is no longer used.

9. The owners of items listed on the register can access financial, technical or other assistance, by agreement with the minister, to ensure a building's conservation. Taxation relief is available for owners of heritage-listed buildings at a 20 cents in the dollar rebate for conservation work.

10. The Heritage Council is made up primarily of ministerial appointments from nominees representing the historical, architectural, planning communities, the National Trust and organized labour.

11. LEPs (and DCPs) and REPs must be displayed for public comment at the draft stage and LEPs must be re-exhibited if they are significantly amended before they will be approved by the minister. SEPPs, on the other hand, are drawn up at the minister's request. Though they *may* be publicized and submissions taken, this is not a legal requirement.

12. For instance the court has recently approved on appeal a development at Darling Point on the western edge of the CBD that exceeds allowable density by 280 per cent and height limits by 33 per cent.

13. These amendments simplified the task of applying for development approval through a 'one-stop-shop' single application to address planning, building and other environmental matters. This provides for speeding up the granting of these additional approvals (e.g. under the Heritage Act) by consolidating the previously sequential process so that all approvals are linked to the development approval.

14. By 1973 the six major life assurance offices held 18 per cent of their assets in property. AMP increased its assets in property from 12 per cent in 1968 to 22 per cent in 1977 (Daly 1987).

15. The 1971 strategic plan was updated in 1974, 1977, 1980 and 1983. The CSPC produced a new strategic plan in 1988.

16. For instance in the 1971 strategic plan, the action plan for the core retail precinct around Pitt Street suggested the pedestrianization of the street to form the Pitt Street Mall. It took until 1987 before this could be achieved due to fractious relations between council, state government and local retailers.

17. The highest density was, however, sometimes higher than intended by the recommended floor-space ratio of the strategic plan as council was overruled by the state planning authority and the Height of Buildings Advisory Committee which effectively operated as consent authority on building higher than 80 feet (see Ashton, 1993).

18. The pro-development ethos that prevailed in both state and local government during these years was demonstrated in two major state-sponsored proposals for extensive redevelopment of low-income residential areas in inner Sydney as high-rise commercial office space; at the Rocks and at Woolloomooloo. While the proposal for the Rocks involved the formation of a special purpose authority, the Sydney Cove Redevelopment Authority, and the removal of local government planning influence, the proposal for Woolloomooloo was supported by Sydney City Council. After long drawn-out resident-activated political opposition to the redevelopment, including Green Bans by the Builders' Labourers' Federation, neither project proceeded. Detailed accounts of the conflict over the Rocks proposals can be found in Morgan (1991, 78–87) and of the Woolloomooloo proposals in Ashton (1993, 103–9) and Fitzgerald (1992, 275–303).

19. The Local Government Act (1993) gave greater responsibility and accountability to local authorities and encouraged partnerships with other levels of government. Meanwhile

Regional Organizations of Councils have taken on significant roles in local economic development.

20. Between 1986 and 1996 finance and business services employment grew further by a massive 78 per cent.

21. Financial reforms under Labor Finance Minister Paul Keating (1983–5) involved floating the dollar, removing capital controls and allowing the entry of foreign banks.

22. Two-thirds of Australia's banking and finance industry, three-quarters of Australia's financial services and two-thirds of the regional head offices of MNCs with Asia Pacific operations (275 in total) are located in Sydney, predominantly in the CBD. Sydney's Futures' Exchange is the largest in the Asia Pacific, its stock exchange the twelfth largest in the world, and it has the most densely concentrated node of fund management activities in Australasia.

23. Traditionally, state planning authorities had encouraged the location of office development in the CBD. Since the property boom of the 1970s, however, developers have also exploited opportunities to develop lower-rent office space in suburban centres across the city and this served the dispersal of office employment from the CBD to Parramatta, North Sydney/St Leonards and Chatswood. These centres had developed as mature business locations and, indeed, absorbed one-third of information service employment expansion between 1981 and 1991 (see Fagan 2000). Economic forecaster B. I. S Shrapnel predicts that by the end of the decade, 25 per cent of Sydney's office stock will be located in the suburbs, mainly in suburban office centres and office parks (O'Connor et al., 2001). Nonetheless, the CBD concentration still dominates with 70 per cent of the office space in the metropolitan area, while North Sydney has 12.5 per cent, Parramatta has 8.5 per cent, Crow's Nest/St Leonards and Chatswood have 5 per cent each. The smallest ever share of office output produced in the inner region in a given year was 57.6 per cent in 1995 (SCC 1999).

24. Between 1991 and 1999 population of the City of Sydney increased from 7,000 to 25,000 and is predicted to reach 40 000 by 2021 (SCC 2000). Asian investors and overseas students contributed strongly to this growth.

25. The Regent, the Intercontinental, the ANA, the Carlton and the Ritz were all constructed in the 1980s.

26. While the CBD dominated 50 per cent of metropolitan retail sales in the 1940s, this fell dramatically to 30 per cent by 1962, 13.6 per cent by 1974, and 10.5 per cent by 1980. The tourist-driven revival of the CBD's retail function has seen its retail share stabilize at around 10 per cent.

27. Including Chanel, Donna Karan, Louis Vuitton, Hermes, Gucci and Giorgio Armani.

28. The demolition of the art deco State Bank in Martin Place to make way for a new State Bank headquarters, and of the Hordern Department Store building, were particularly significant losses.

29. Sartor has since held power for a remarkable 12 years since 1991.

30. Vast council resources have also been spent on improving public spaces and thoroughfares in the city. Support for this public spending was forthcoming because of Sydney's pending role as host of the 2000 Olympics. The public space improvement programme, however, also involved levering significant private funding from CBD-based corporate interests. Thus it was an instance of council entrepreneurialism rather than merely subsidization.

31. For instance, Lend Lease has developed Jacksons Landing, a self-contained urban village and the largest redevelopment project in Australia.

32. Just to the west of the CBD, the Rocks is the location of the original European colonial settlement and is rich with heritage buildings which were threatened by the SCRA's plans to redevelop the area as high-rise commercial office space. After a long-drawn-out campaign of public resistance, much of the Rocks' heritage has been conserved through the area's redevelopment not as office space but as one of Sydney's most visited heritage tourist attractions. Planning control, however, has never returned to the local council.

33. The A$800m Better Cities programme ran from 1991–6 as a rare instance of direct Federal intervention in urban policy. It aimed to develop a series of projects to demonstrate 'how higher-density, planned urban development integrating housing, services and employment will help the three aims of economic efficiency, environmental sustainability and social justice' (Orchard, 1995, 71).

34. A similar example can be seen in the case of SEPP No. 47 – Moore Park Showground (1995) which was used to enable the redevelopment of the Moore Park Showground for film and television studios and film-related entertainment facilities, again overriding the LEP and significant local council and community opposition. Once again, the Minister was position as consent authority.

35. The development approved by government came from Walsh Bay Properties, a specially formed conglomerate of large-scale residential developer Mirvac and transport development specialist Transfield.

6
Dublin: property development and planning in an entrepreneurial city

Andrew MacLaran and Brendan Williams

6.1 Introduction

Modern urban planning in Ireland dates from 1963. This rather late start is largely attributable to the country's having gone through a period of virtually unrelieved economic stagnation during the preceding 150 years. Only with economic growth in the 1960s, the first Irish 'economic miracle', did planning become a necessity. Demand for new buildings of every type led to a restructuring of the urban environment, with land uses being upgraded in the existing city and rapid growth taking place at the periphery.

Geographical expansion facilitated the continuing onslaught on the city's insanitary and densely occupied slums. As inner-city population densities fell, a shift in the residential location of the workforce took place. The development of suburban industrial estates tapped into this suburbanizing labour force, with manufacturing and wholesaling operations relocating from congested inner-city environments to more readily accessible new premises. Planners encouraged this process, permitting a reorganization of inner-city functions. However, the relocation of industrial plants from the inner city and shrinking employment in the port hit the employment base of the predominantly blue-collar communities. Entry into the European Economic Community in 1973 brought intensified competition for Irish industry, while the 1980s brought economic stagnation. By the late 1980s, inner-city unemployment rates exceeded 35 per cent.

These changes created potentially profitable opportunities for developers to reconfigure inner-city land uses to more profitable functions. In particular, the rising demand for office space generated enormous pressure to upgrade land use. From the 1960s, prestigious parts of the eighteenth-century city became transformed from residential to office functions (see MacLaran, 1993). In what was developing as the city's office core, large eighteenth-century houses were either converted to office use or demolished to make way for larger, modern space-efficient buildings which would maximize the rent-generating capacity of their sites. In contrast, in less desirable secondary areas adjacent to the office core, slum clearance, industrial relocation and business closures created a landscape of building abandonment and dereliction resulting from a lack of user demand for space.

By the mid-1980s, the logic of the private market and the geography of development profitability (see Chapter 2) had become dramatically inscribed in the urban landscape.

While buildings in prestigious inner-city areas were under almost permanent threat from redevelopment, others in less salubrious localities were in danger of falling down through neglect. Then, during 1986, in an effort to boost employment in the ailing construction sector, the Irish government established a series of property-based urban regeneration programmes. As property development activity increased, the depression of the 1980s slowly gave way in the 1990s to unprecedented economic growth; the so-called 'Celtic Tiger'. Growth was driven by high levels of capital investment by foreign companies and strong export performance in high-tech sectors such as electronics, computer software, chemicals and pharmaceuticals, combined with rapid employment growth in the services sector. Between 1993 and 1996, the Gross Domestic Product (GDP) increased by 31 per cent, continuing thereafter to grow at over 8 per cent annually until the end of the century. Whereas Irish GDP per capita had been only 63 per cent of that of the UK in 1987, it had surpassed it by 2000. Unemployment nationally dropped from 18 per cent to around 4 per cent and the long-run tide of emigration turned into a net influx of 18–20,000 persons annually.

Nowhere were the consequences of economic development in the second half of the twentieth century more sharply felt than in Dublin, its landscape becoming transformed in response to the increased demand for accommodation of all types. Between 1960 and 2000, over 262,000 new dwellings were constructed in the capital as its population rose from 718,000 to over a million residents. Over 3 million sq. m. (32 million sq. ft.) of industrial space, 2.1 million sq. m. (22 million sq. ft.) of offices and almost 500,000 sq. m. (5.4 million sq. ft.) of floor space in shopping centres were developed. Yet, during this period of rapid metropolitan growth and transformation, the urban planning system responded poorly to development forces. Indeed, in key regeneration areas, its role became marginalized during the 1980s as a result of urban revitalization initiatives devised by the national government. Moreover, it failed to act strategically to ensure that the planning of land-uses and the development of transportation infrastructure were co-ordinated, thereby storing up immense problems for the future which will be enormously costly to address.

This chapter reviews the deficiencies of traditional Irish urban planning approaches which prevailed for over 25 years; a system that long proved inadequate in controlling the destructive impact on the urban landscape of the private-sector property development forces. Simultaneously, it also failed to generate development in those areas where it was required. As its inadequacies became increasingly apparent, the national government increasingly demanded local authorities adopt a 'can do' entrepreneurial culture and sought to promote a more 'proactive' agenda in which urban planning was increasingly to become overtly facilitative of private-sector property development.

Marginalized by the creation of special-purpose development agencies set up to promote development, unconvinced of the merits of the new entrepreneurialism and reluctant to discard their traditional roles of watchdogs for the 'public interest', urban planners sought renewed relevance based on local-area planning. Simultaneously, however, major planning problems of strategic metropolitan significance became increasingly apparent, to which the system has yet to respond in a satisfactory manner.

6.2 The planning context

The 1963 Planning Act, which established the modern planning system in Ireland, confirmed local authorities as the appropriate planning bodies. Planning in Dublin therefore falls within the remit of four local authorities. Dublin Corporation (renamed Dublin City Council in 2002) administers Dublin County Borough, comprising the central city, the inner suburbs and much of the northern urban fringe. Fingal County Council, South Dublin County Council and Dun Laoghaire-Rathdown County Council each comprises an element of outer suburbia together with adjacent rural districts. Until the early 1990s, some degree of co-operation between the metropolitan authorities had been ensured through the most senior urban manager acting as City and County Manager. Similarly, the most senior planner was Chief Planner for the city and surrounding county areas, though not Dun Laoghaire Borough. However, both these positions were terminated during local government reform in the early 1990s and administrative co-ordination now relies on a Regional Authority, whose impact has to date been very limited.

The basis of urban planning lies in the zoning of land for different categories of use, together with development control. In addition, there is provision for the listing of buildings for protection against unapproved alteration or demolition.

6.2.1 Development plans

Each local authority is charged with adopting a development plan. This comprises written statements and maps, setting out planning goals and covering such issues as the zoning of land-uses, traffic and infrastructural developments, areas of obsolescence and renewal and the preservation and enhancement of amenities. It was intended that development plans would be reviewed every five years. In practice, the process of review proved cumbersome and took far longer, though the power of elected councillors to vary aspects of the plan according to changing circumstances has mitigated the consequences of such delays.

In drawing up development plans, public participation is provided for, both directly and through elected councillors. The plan, devised by local authority planning departments, is commented upon by the elected members of the authority who may make variations prior to adopting it in draft form. The Draft Plan is then placed on public display, inviting comments and representations. Planners review the comments and appropriate amendments are made. The amendments are then exhibited for public display and comment. The councillors are responsible for adopting the Development Plan, while the government minister reviews the plans of adjacent authorities to ensure that they do not conflict on major matters (see MacLaran, 1993).

In recognition of the degree of dependency of planning goals on private-sector operations, it has been the practice of Dublin Corporation to survey businesses in the city to encourage their input into the process of Development Plan formulation. Comments by businesses are also likely to be made at an earlier more influential stage, prior to their formulation (McGuirk, 1991). Effective participation by the public therefore tended to be less effective, coming at a later stage. Furthermore, participation was sometimes hampered by planners' attitudes, regarding themselves as representing the 'common good' or 'public interest'. One commented to McGuirk (1991,266) that 'we are employed to do the job for the public. [Participation] is like buying a dog and barking yourself!'

6.2.2 Development control

Developers apply to the local-authority planning department for permission to develop. 'Developments' include new constructions and extensions above a certain size, changes to building or land-use functions (e.g. from residential to office). Planners ensure compliance with designated land-use zonings and control the scale of development by regulating plot density and building height. They also vet all aspects of the architectural design. These development controls can represent a significant obstacle and developers often decry the resultant delays.

Refusals can be appealed by the applicant to higher authority, originally the relevant government minister and, since 1977, to a planning appeals board, An Bord Pleanala (see MacLaran, 1993). A right of appeal also exists for third parties, irrespective of whether or not they possess a direct interest in the proposal. This allows broader environmental and social issues to be considered but has been a source of recurrent complaint by developers over the delays resulting from appeals sometimes regarded as pernicious. However, major limitations exist to effective public participation (see McGuirk, 1991). Inevitably, community groups have varying financial resources, expertise, knowledge and confidence, influencing the impact of their involvement. But all community groups tend to be financially disadvantaged compared to business and development interests. Moreover, as developers hold the initiative and often enter pre-application negotiations with the planners, the public only learn of proposals much later. While not binding on the planning department, such informal discussions enable developers to assess what the planners are likely to permit and where best to apply political pressure.

6.3 Limitations of traditional urban planning in Dublin

6.3.1 Constitutionality and constraint

Private property rights are enshrined in the Irish constitution. Although these rights are tempered by considerations of 'social justice' and 'the exigencies of the common good', when regulating private-sector development, urban planning has laboured under a fear that certain actions might face legal challenge. Such considerations underlay the failure of national politicians to act on the recommendations of the Commission of Enquiry into the costs of development land which reported in 1973. This favoured the taking into public ownership of designated development land at the edge of the urban area at current (agricultural) use values plus a disturbance allowance.

6.3.2 Contextual isolation

A major shortcoming of Irish urban planning has been its almost complete separation from other national, regional, economic or physical planning processes. It has thus lacked strategic significance. Individual planning departments may even pursue policies which run counter to those of their neighbours and this has become particularly serious in the Dublin commuter belt.

These contextual shortcomings have only recently been addressed. Strategic Planning Guidelines for the development of the Greater Dublin Area (GDA), an area including the commuter zone lying in the surrounding rural counties, have been devised by private-sector planning consultants commissioned by the Dublin Regional Authority, the rather powerless body established to effect planning co-ordination in the GDA. The first Strategic Planning Guidelines were published in 1999, a review and update being published in 2000 as initial population and growth projections had already been exceeded. While generally welcomed as a contribution to resolving Dublin's urban development problems, the absence of serious funding commitments or proposals for implementation reduce the guidelines' credibility. Moreover, the fact the guidelines were brought forward after six of the affected local authorities had already adopted or were in the process of adopting their Statutory Development Plans also created issues for conflict.

The guidelines emphasized sustainable integrated urban development along existing or projected transportation corridors separated by green wedges where development would be reserved for local needs. However, some local authorities have already forwarded planning proposals contrary to the guidelines. Contentious proposals include the development of major business parks at the Dublin border in County Meath and near Bray, County Wicklow. In June 2001, the Planning Appeals Board rejected plans for a business park in Wicklow which had been endorsed by the local planning authority, signalling the board's efforts to uphold the guidelines on the basis of 'proper planning and sustainable development' criteria (*Irish Times*, 14/06/2001).

The policy shift towards sustaining current economic development is also evident in the creation of a National Development Plan for the Period 2000–6. These proposals aim particularly to deal with the national requirement for infrastructural development, with particular emphasis on public transport in Dublin. With a funding requirement over €50 billion, this represents the initiation of an intense public capital investment programme aimed at improving the competitiveness and efficiency of the Dublin region.

A National Spatial Strategy is also currently being devised and debated and the announcements of key decisions on such policy is expected at the end of 2002. Analysis of the proposals indicates that decisions on several major areas with particular relevance to the Dublin Region are well advanced. The aspiration to distribute growth more widely in Ireland has been reflected in reduced grant aid to businesses locating in Dublin. Attempts to redirect development will have investment consequences for planning and development of the Dublin region. However, redirecting urban growth to alternative areas possessing the capacity to absorb growth may prove difficult.

Incredibly, given the intimacy of their relationship, land-use planning and transport planning have long been undertaken completely separately. However, new institutional arrangements have recently been proposed by the government to facilitate the better co-ordination of land use and transport planning (Department of the Environment and Local Government & the Department of Public Enterprise, 2001). Nevertheless, plans to invest €22 billion in the development of transport infrastructure are already well advanced, prior to putting such co-ordinating arrangements in place. Indeed, some elements (a light rail system in Dublin and a road tunnel to connect Dublin port and the orbital motorway) are already under construction.

Moreover, on a more general level, the problematic spiral of infrastructure-related spending on transport being followed by further commercial property development seems not yet to have been recognized in planning circles. Briefly, any improvement in radial transport systems which increases the accessibility of the city centre enhances the development value of central sites. This creates pressure for land-use intensification, with high-value (e.g. office) functions replacing lower-value (industrial and residential) functions. The new developments are then normally staffed by suburban commuters, contributing to renewed congestion and generating yet further demands for transport improvements. Yet additional transport-related improvements will further increase accessibility, bring about increased land development pressures and more land-use intensification and commuting Indeed, most cities have failed to address this problem.

6.3.3 The permissory basis of the system

The twin elements of land-use zoning and development control embodied in the 1963 Act created an urban planning system that was essentially permissory. Planners could approve or refuse development applications. At times of development boom, planners tended to become overwhelmed with the number of applications lodged. However, at times of development inactivity when few applications were forthcoming, the power of planning was highly limited.

Although the Act had envisaged a more active involvement of local authorities in development activities, empowering them to secure land for development or develop it directly, local authorities were never provided with adequate finance to permit such activity. This ensured that these powers could never be seriously used other than for the development of roads, social housing and parks.

6.3.4 Political pressure

Planners have also frequently come under pressure from local authority councillors and members of the Dail (parliament), lobbying on behalf of constituents. Pressure may also be exerted by central government through the Department of the Environment (see McGuirk, 1991, 243). With a simultaneous collapse of all major property sectors in Dublin after 1982, planners were entreated to grant planning permission for almost any development in order simply to create construction jobs. A circular in 1982 from the Department of the Environment advised planners that development should only be refused where there were serious objections on important planning grounds.

As elected councillors may revoke or modify a permission to develop land and require the city manager to grant planning permission resulting in a material contravention of a development plan, corrupt influence can be brought to bear over applications for land re-zoning. Such matters are currently being investigated by a tribunal of enquiry into corruption in the planning system involving payments made to senior local government officers and politicians for favourable planning decisions and land re-zoning. The tribunal has also exposed the extent to which the politicians have subverted the planning system for party or personal benefit (see McDonald, 2000).

6.3.5 Planning by prescriptive formulae

It has long been an unfortunate tendency for developers to build to the minimum quality criteria set down by planning. This is especially apparent in private-sector residential developments. Prescriptive formulae cover a wide range of elements from housing set-back lines to the quantity of open space, the width of roads and maximum permitted densities. Rather than creating intimate residential environments, estates were organized around road plans designed to hasten traffic flow (Figure 6.1). Mortgage lenders also favoured standard housing designs, easy to sell in the case of default. The outcome was a suburban environment characterized by an overpowering monotony, comprising predominantly of two-storey detached, semi-detached or terraced houses with gardens, built at 20 to 25 houses per hectare with a mandatory

Figure 6.1 *Residential development reflecting prescriptive planning codes, developers' profitability criteria and the conservative preference of mortgage lenders for 'standard' properties*

provision of 10 per cent open space. The low densities proved extravagant in their consumption of land and costly for the provision of services such as refuse collection and public transport.

6.4 Developers and investors

Dublin's property developers have comprised a very heterogeneous group. They include individuals and partnerships, private and publicly quoted companies, the development departments of financial institutions, particularly the life assurance companies, and public-sector bodies. Most developers have been Irish-based but they have been joined since the 1960s by several from overseas. These have been predominantly of UK origin, including MEPC and British Land both of which have been responsible for developing office and retailing schemes, and the former Lyon Group which developed industrial units. Some developers have tended to specialize in one particular sector, residential developers being most likely to specialize. However, even they have entered other sectors, several having developed office schemes recently.

Even for a single property sector, such as offices, the range of developers has been large. It includes developments undertaken by owner occupiers, ranging from small accountancy and legal firms to the corporate headquarters buildings of banks and insurance companies. There are also owners of sites who decide to redevelop them for a profit, but for whom property development is not a core element of normal business. In addition, development for owner occupation is frequently undertaken by the public sector, including the semi-state (nationalized) industries such as the Electricity Supply Board and Aer Lingus, by local authorities and the Office of Public Works which develops offices for state operations.

A second group of office developers includes those for whom property development is a continuous part of their operations. Commercial developers range from individual operators and private companies (e.g. Hardwicke, John Byrne Group, Clancourt, Duke House, Treasury Holdings, and Durkan Bros.), to companies quoted on the stock exchanges either in Ireland or abroad (e.g. Abbey, Green Property and Dunloe).

These commercial developers have been joined by the development operations of financial institutions, notably the insurance and life companies such as Irish Life, New Ireland and Hibernian Insurance. Even during the office boom of the 1960s, when few Irish institutions had become actively involved in speculative office development, UK-based institutions such as Royal Insurance, Guardian Royal Exchange, Scottish Legal, Standard Life, Friends Provident and Norwich Union began developing office schemes in Dublin. However, the closing years of the twentieth century saw a growing tendency for the institutions to withdraw from development, reverting to their more traditional role as purchasers of completed schemes.

Finally, in order to gain greater direct access to development profits, construction companies have also undertaken office development directly or have done so through setting up development subsidiaries.

As a target for international investment capital, the Irish property sector is small-scale and dominated by Dublin. However, even here, the stock of modern office property amounts to less than 2.1 million sq. m. (22.5 million sq. ft.). Most investment demand for Irish property has been local in origin and, until the late 1980s when controls over investing Irish funds overseas were terminated, was driven by a desire for portfolio diversification in a context where the share market was very limited. The major investors were the life offices (e.g. Irish Life, New Ireland, etc.), Irish pension funds, acting directly or through the Irish Pension Fund Property Unit Trust, and the investment banks. The property sector has nevertheless attracted overseas commercial developer-investors such as British Land and MEPC, while institutional developer-investors have included Royal Insurance, Friends Provident and Norwich Union. The UK National Coal Board pension fund also acquired property in Dublin as did the National Mutual Life Association of Australasia.

6.5 Development activity

Each of the major property sectors has been characterized by phases of boom and slump in development activity and, with each phase, have come changes in the geography of development. This is most apparent in the offices sector. During times of development boom, activity has expanded into secondary areas adjacent to the core, while retrenchment into prime areas occured during slumps. However, in the most recent office development boom of the late 1990s, considerable development has been deflected to the outer suburbs (MacLaran and O'Connell, 2001; MacLaran and Killen, 2002). To some extent, this shift reflected changing demand. One aspect has been associated with the creation of 'call centres' engaged in teleservices operations. These range from IT-related sales and after-sales backup for companies such as IBM, Dell and, formerly, Gateway, to hotel and car-hire reservation services of Best Western and Hertz. With European-wide operations and dependency on telecommunication systems, such offices have wide freedom of locational choice. Cheap suburban offices were therefore favoured when large amounts of floor space were required. In a city marred by worsening traffic congestion, this also offered the prospect of reduced commuting times for employees.

Urban planning restrictions also played a key role in office suburbanization. The increasingly restrictive pro-conservation planning ethos of the Corporation planners, combined with competition for city-centre sites from other functions, such as hotels and high-priced residential uses, reduced the number of sites available for office development in the core. Simultaneously, the more liberal planning regimes of the suburban local authorities, eager to gain the income from commercial rates (property taxes) after the abolition of domestic residential rates in the late 1970s, encouraged office developers to widen their operational fields (Bertz, 2002). This resulted in a significant change in the location of office development. Over 85 per cent of the office space developed between 1960 and the end of 1989 had been built in the prime office core (Dublin 2 and 4), with less than 10 per cent being located in the outer suburbs. However, from 1990 to 2000, during the greatest ever office development boom in the city, only 27 per cent of new floor space was developed in the core, whereas 53 per cent was located in the outer suburbs. Unwittingly, Dublin had started to create an 'Edge City'.

6.6 Urban planning and development

By the mid-1980s, the impact of property development on the central city had been considerable. However, it was becoming apparent that the urban planning system had serious limitations and sometimes even exacerbated the problems. For example, by zoning too large an area of the central city as suitable for commercial development, planners unwittingly encouraged long-term speculative site assembly for commercial redevelopment over wide areas fringing the core, bringing widespread dilapidation and often causing the displacement of blue-collar employment.

6.6.1 Development scale

Although most office developments in Dublin have been small-scale, fewer than 30 of its 760 modern office buildings being over 10,000 sq. m., the preference for larger schemes which can achieve development economies has nevertheless had an impact on the urban landscape. This is evident around St. Stephen's Green where the eighteenth-century townscape has undergone radical transformation. To the north of the river Liffey, the Salvation Army building remains as solitary witness to the grain of historical development, now dwarfed by the Irish Life Centre, a mixed retail and office scheme (Figure 6.2). Built on a former builders' merchant's yard and benefiting from ease of site assembly, the very scale of the scheme has permitted it to turn its back on the local

Figure 6.2 *Dwarfed by the scale of its Irish Life Centre neighbour, the Salvation Army building (bottom left) preserves the more intimate scale of the historic townscape*

environment and draw the zone of prime office rentals over the river to encapsulate the development. Reflecting its intrusion into such a secondary location, it was equipped with a dry moat and retractable drawbridges, affirming its fortress-like character.

6.6.2 Site assembly and dereliction

The impact on Dublin's streetscapes of widespread redevelopment was compounded by the lengthy time required for site assembly in the city centre (see Chapter 2). Subsequently, the need to await an upturn in user demand might occasion yet further delay in building. Nowhere was this better exemplified than on Leeson Street (see Chapter 2, Figure 2.8) where eighteenth-century buildings remained derelict for a decade prior to redevelopment. Indeed, the assembly process at one site on St. Stephen's Green was so lengthy that it contributed to the demise of the developer, the scheme eventually being completed by another developer for retailing rather than as offices.

6.6.3 Plot density and displacement

Urban planning has been largely successful in controlling building height and density of development, thereby preserving Dublin as a low-rise city, built at an intimate 'human' scale. Displacement problems nevertheless arose. A low density of permitted commercial development, with plot ratios of only 2.5 : 1 was generally enforced. This created a wave of office property development pressure pushing outwards into areas adjacent to the office core. It was especially damaging for buildings located in the prestigious eighteenth-century areas to the south and east of the core where developers were for too long able to secure permission for redevelopment.

6.6.4 Building protection

Planners long proved unable to protect the valuable heritage of eighteenth-century buildings, streets and squares developed during Dublin's 'Golden Age' (see MacLaran,

Figure 6.3 *Mount Street Lower. Redevelopment has caused the loss of numerous eighteenth-century buildings in Dublin's office core*

1993). Not until the 1990s were any interiors listed for protection. An Taisce (the National Trust) estimated that between 1980 and 1985, some 80 listed buildings were demolished or permission had been granted for their demolition, with another 50 suffering unauthorized alterations detrimental to their architectural character (An Taisce, 1985). While eighteenth-century buildings in the office core were in constant threat from redevelopment, in secondary areas the paucity of user demand led to widespread dereliction and the danger of buildings falling down (see MacLaran, 1993). Sometimes, in the absence of permission for demolition, owners might remove lead flashing and roofing slates to allow rain to penetrate, eventually rendering the building structurally unsound. Its compulsory demolition would then be required by the dangerous buildings inspectors, working for the same local authority whose planners were trying to preserve it.

Figure 6.4 *Mountjoy Square (1993). Lack of user-demand may lead to the suspension of building maintenance and to dereliction*

Thus, from an origin bursting with high ideals and enthusiasm, by the 1980s the planning system had degenerated into a bureaucratic procedure, held in disregard by developers and community groups alike. Developers viewed it as a costly and unnecessary bureaucratic obstacle, while community groups found it difficult to penetrate and unresponsive to local needs (McGuirk, 1991). Moreover, in the mid-1980s, it was becoming increasingly recognized in government circles that Dublin Corporation, and the planning system in particular, lacked the ability to tackle the scale of Dublin's problems; a landscape which presented a facade of inner-city dereliction and social problems associated with high unemployment.

6.7 Towards interventionist approaches in planning

Tentative attempts by planners to influence the geography of office development more actively had been attempted during the office boom of the 1970s. With the aim of guiding development to the north side of the river Liffey where urban renewal was much needed, planners offered higher than normal plot ratios for developments. However, this had a very limited impact because of the conservative locational preferences of office establishments. In the 1970s, an attempt by the Corporation to become more actively engaged in development involved its entering a partnership with Irish Life Assurance for the development of the ILAC Shopping Centre. However, the experience left much to be desired. Irish Life found the naïve aspirations, lack of understanding of development economics, absence of any sense of urgency and the Corporation's cumbersome bureaucratic decision-making difficult to tolerate (MacLaran, 1993). If future joint ventures were to have any prospect of success, they clearly had to be based on a different footing.

Towards the end of the 1970s, growing recognition that the central city faced serious difficulties and policy deficiencies led to the establishment of a government inter-departmental committee and an Inner City Task Force. The problems were highlighted in an influential report (Bannon et al., 1981). The 1982 general election presented an inner-city Independent community candidate with the balance of power in the formation of the incoming government. Negotiations for his support centred around a complex inner-city development package, including initiatives ranging from inner-city housing projects and economic development, embodied in the Urban Development Areas Bill, 1982. While never enacted, it provided the key elements of subsequent renewal initiatives.

Lacking confidence in the ability of Dublin Corporation to respond to the depth of the inner-city problems, particularly the existence of over 600 cleared sites and derelict buildings totalling 65 ha. (160 acres), the national government responded in 1986 by devising a series of property-based regeneration initiatives. These provided financial inducements to developers, investors, owner occupiers and tenants of properties in defined renewal areas (MacLaran, 1993; Davis and Prendergast, 1995). The 'Designated Area' policies had a major impact on the geography of property development in Dublin over the following decade.

Simultaneously, the operational climate of the Corporation departments became transformed by a managerial ethos concerned increasingly with city boosterism, promoting and selling the city in order to attract international investment. The limited discussion and consultation which had taken place between central government ministries and local government departments gave emphasis to the marginal role which was envisaged for Corporation planners during a period of unprecedented urban development (McGuirk and MacLaran, 2001). Indeed, to circumvent traditional urban planning entirely, special purpose development agencies were established for key sites.

6.7.1 Designated areas

The Urban Renewal Act (1986) and the Finance Act (1986), provided for the establishment of 'designated areas' of Irish cities where tax incentives would be made available to promote the development, refurbishment and occupation of properties in districts requiring renewal (see MacLaran, 1993; Williams and MacLaran, 1996). The aim was to stimulate employment in the construction industry, then experiencing an unemployment rate of over 45 per cent following the collapse in the office, industrial and residential sectors. The incentives included:

- Taxation allowances for investors in respect of expenditure of a capital nature for the construction or reconstruction of commercial buildings. Such costs could be set against income or corporation (company) tax.
- For tax purposes, occupiers of new or refurbished properties were granted an allowance amounting to twice the value of rent paid.
- Remission of commercial rates for a 10-year period on a sliding scale, relating to the erection, enlargement or improvement of commercial premises.
- Income-tax relief for owner occupiers' expenditure relating to newly constructed or refurbished private dwellings.
- For tax purposes, investors in rented residential properties within specified size limits could offset construction costs of such units against rental income.

Their impact was to increase substantially the potential redevelopment value of inner-city sites. Property owners who initially failed to benefit because their sites were located on the 'wrong' side of the incentives boundary, commenced a lengthy process of lobbying of the government minister, urging the geographical extension of the designated zones. Expansion duly occurred between 1988 and 1990, by which time the incentive areas covered a large proportion of the central city outside the main shopping streets and prime office area (see Figure 6.5). Among the original 'urban renewal' designations was a 48-ha. (120-acre) green field site in the peripheral new town of Tallaght, included to facilitate the development of a new shopping centre to service the growing population.

6.7.2 Docklands

Provisions were also put in place to effect the redevelopment of the Custom House Docks (Figure 6.6), a redundant dock site close to the city's central business area which the government wished to see redeveloped (see Gahan, 1993). Here, the government decided to control its planning and redevelopment by establishing a special agency. The Custom House Docks Development Authority (CHDDA) was endowed with full planning powers and charged with overseeing and promoting redevelopment. An integrated development project was envisaged to include a wide range of business, residential and recreational functions. Companies could avail of 100 per cent capital allowances for business premises (54 per cent in the first year and 4 per cent thereafter), while residential landlords could offset the allowable cost of premises against their taxable income. Plans for an International Financial Services Centre (IFSC) were added, involving enhanced incentives for commercial development and a reduced corporation tax rate of 10 per cent for companies involved in internationally traded services.

The CHDDA held an open competition among developers interested in participating in the redevelopment and, by mid-1987, it had devised a general planning scheme to

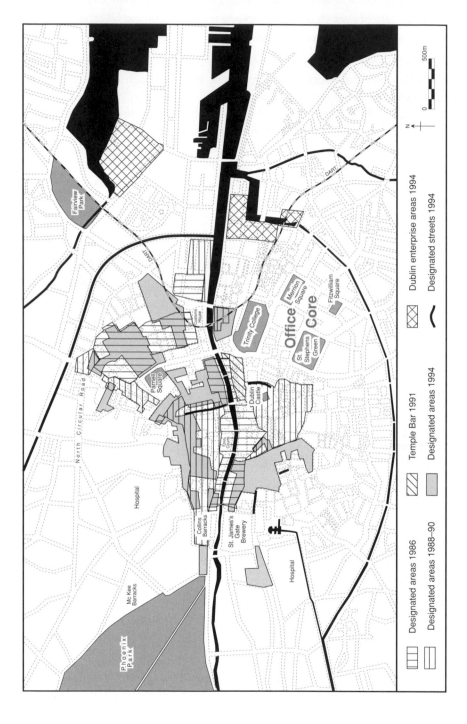

Figure 6.5 *Designated Areas. Incentives for property-based renewal were spasmodically extended to cover much of the city centre*

Figure 6.6 *The Custom House Docks, 1986 (© Peter Barrow, Dublin)*

Figure 6.7 *The International Financial Services Centre and associated developments at the Custom House Dock, 2002 (© Peter Barrow, Dublin)*

attract interest. The winning submission, evaluated on the basis of its design, financial backing and 'deliverability', was lodged by a British and Irish consortium comprising British Land, Hardwicke and McInerney Properties, all of which had considerable development experience in the city. The scheme involved a total of more than 70,000 sq. m. (750,000 sq. ft.) of office space. It also provided for a 300-bedroom hotel, a 5,000-seat conference centre, 200 residential apartments, museums, retailing elements, restaurants and pubs, an entertainment centre, community and training space and underground parking. After signing the Master Project Agreement, the land was passed to the development consortium, trading as the Custom House Docks Development Company, on a 200-year leasehold as individual components of the development were undertaken.

The geographical remit of the CHDDA was extended from the original 11 ha. (27 acres) site to over 20 ha. (50 acres) and in 1997, on its being succeeded by the Dublin Docklands Development Authority (DDDA), this was increased to embrace 526 ha. (1,300 acres) of decaying docklands which had suffered from the physical and socio-economic decline typical of de-industrialized ports (Drudy, 1999). Although the DDDA did not obtain the same sweeping planning powers as in the Custom House Docks, it was empowered to develop special detailed planning schemes for areas thought to be in need of particular redevelopment assistance and these are largely exempted from local authority planning control.

6.7.3 Temple Bar

In 1991, the government made special provision for the renewal of Temple Bar. This riverside district comprising over 80 ha. (197 acres) at the heart of the city, whose development could be traced back to the seventeenth century and where there still existed buildings dating back to the 1720s, was situated immediately south of the river Liffey between the city's two main retailing streets. It had suffered from blight and decay resulting from long-term plans to redevelop a large site as a bus station. Cheap premises on short leases and the proximity to the central area had created a 'left-bank' atmosphere of hotels, pubs, cheap cafés and restaurants, theatres, galleries, recording studios, second-hand goods and clothing stores. Nowhere else in the city was there to be found such a heterogeneity of functions. The government's aim was to create a cultural, artistic and tourist quarter by protecting and enhancing the types of function which had already apppeared (see MacLaran, 1993).

Legislation established two companies to oversee rehabilitation. Although the Corporation retained its formal planning control, this initiative paid little heed to the city planners' existing detailed 'action plan' for the area and the role of the Corporation planners was effectively marginalized (Montgomery, 1995).

Tax incentives provided for the purchase price (net of land cost) of an appropriately refurbished residence to be allowable against the income tax of an owner-occupier. Unlike elsewhere in the designated areas where tax incentives were available for all property-related developments, incentives for businesses in Temple Bar were made available only for buildings accommodating functions approved by Temple Bar Renewal Ltd, the company charged with creating the required functional mix. The second company, Temple Bar Properties Ltd, acted as a development company for the area. It took over the publicly-owned properties, primarily the portfolio of the transport authority which had been assembling sites for the proposed bus station, and immediately embarked on a programme of refurbishment and infill development.

6.7.4 Enterprise areas

In 1994, the government established two enterprise areas adjacent to the docklands, fiscal incentives being made available to promote industrial operations. However, it was the Industrial Development Authority (IDA), rather than the Corporation, which was charged with securing and approving appropriate companies, again limiting the role of local planners. The IDA adopted a broad interpretation of the functions deemed appropriate for support, with a wide range of activities being attracted, including software companies, teleservices and research, most of which were office-based.

6.8 Development impact of the initiatives

With such a variety of schemes, initiatives and development agencies operating concurrently, isolating the effectiveness of individual programmes is difficult. However, there is no doubt that the package of incentives had a major impact on the inner city by sparking a development boom in the office and residential sectors (MacLaran, 1993; Williams and MacLaran, 1996).

The timing of the introduction of the Designated Area incentives was fortuitous, occurring just as the offices sector was about to enter a development boom after six years of depressed economic conditions. Accelerating economic growth in 1988 created increasing demand for office space. The office vacancy rate fell to 5 per cent and rentals increased for the few available new buildings. Within a year, the vacancy rate had fallen to 3.6 per cent and rents, which had been static at Ir£107 per sq. m. (£10 per sq. ft.), surpassed Ir£161 per sq. m. (Ir£15 per sq. ft.) for prime new floor space. This provided a sharp impetus to office development activity.

Development was further fuelled by a sudden influx of property investment funds, amounting to Ir£115M in 1989 alone, as investment managers sought to participate in the rising real returns available from the offices sector. As investors chased the limited supply of new properties, initial yields strengthened from 6.5 per cent to below 6 per cent during the year. Developers reacted swiftly to the changed circumstances in the accommodation and property investment markets and construction began at over half the sites where office planning permissions had not yet expired. The government also required the Corporation to release for private-sector development much of the land which it had assembled for its own purposes, such as for social housing. Access to sites with clear title to development enabled the property sector to respond quickly and development got underway immediately.

The completion of new office space leapt from its mid-decade annual norm of less than 25,000 sq. m. (270,000 sq. ft.) to over 80,000 sq. m. (860,000 sq. ft.) in 1990 (see Figure 2.27). A further 120,000 sq. m. (1.3 million sq. ft.) reached completion during the following year. Within two years, the stock of modern office space in Dublin had expanded by over 18 per cent. Over a quarter of this new development, amounting to 60,000 sq. m. (645,000 sq. ft.), was located in the Designated Areas, including the IFSC.

But by the early 1990s, completions had outstripped the scale of demand from prospective occupiers. City-wide office vacancy rates soared, reaching 11 per cent in late 1991, when 120,770 sq. m. (1.3 million sq. ft.) lay empty, 60 per cent of which was in new buildings. Not surprisingly, developers quickly withdrew from further speculative developments and, by 1992, output had fallen to just 16,900 sq. m. (182,000 sq. ft.). Moreover, apart from the special case of the IFSC, office-based businesses proved reluctant to move to non-traditional locations in the Designated Areas. As a result, in the Designated Areas outside the IFSC, over 40 per cent of the newly developed space lay vacant in mid-1992. Inner-city sites in such locations increasingly became considered for alternative functions, primarily residential development. Nevertheless, office development proceeded apace in the IFSC and the Enterprise Areas throughout the 1990s.

The developments in the Designated Areas and Enterprise Areas were undertaken either by owner-occupiers and site owners benefiting from the provision of the incentives, or by commercial property developers who recognized at an early stage the potential of the transformed geography of profitability in the city. By late 2000, some 317,535 sq. m. (3.4 million sq. ft.) of new office space had been developed in the city's incentive areas, comprising 12.5 per cent of the total completed since 1988. Over 54,390 sq. m. (585,000 sq. ft.) were located incentive areas in the outer suburbs. The IFSC accounted for a further 143,265 sq. m. (1.54 million sq. ft.), the Enterprise Areas another

Figure 6.8 *Christchurch Place. Small-scale office developments characterized the initial phase of property-led regeneration in the Designated Areas outside the Custom House Dock*

53,825 sq. m. (580,000 sq. ft.), with the remaining inner-city Designated Areas totalling 66,050 sq. m. (710,000 sq. ft.). The power of public subsidy had clearly demonstrated its capacity to influence the operations of the private development sector.

After the initial office development impact of the incentives, within the Designated Areas the focus switched to residential schemes, a change in emphasis which mimicked that which was simultaneously occurring in both Auckland and Sydney. In the 1980s, a limited amount of private-sector residential development had been attracted to inner Dublin locations as a result of incentives to promote the private renting of dwellings. These allowed landlords to set against taxable income the costs (net of the land element) of acquiring properties for rent. The immediate effect was to engender a surge of apartment construction in Dublin's prestigious inner suburbs. By the late 1980s, a few tentative schemes were developed in inner-city locations. In 1991, the availability of these reliefs became restricted to the Designated Areas, where generous tax relief was also made available to owner occupiers. A residential development boom was generated in the inner city where little private-sector residential building had taken place at all during the twentieth century.

Figure 6.9 *Arran Quay. Oversupply of offices encouraged the development of apartments in the 1990s*

Proximity to the central area became a strong marketing feature and the schemes sold well, not only to landlords but to young middle-class owner occupiers. By early 1997, around 6,000 dwellings, mostly apartments, had been developed in the Designated Areas, with a further 2,700 units being developed on inner-city sites lacking these incentives (McGuirk and MacLaran, 2001). This residential development, geographically more widespread than the initial office developments which had resulted from the incentives, virtually eliminated the dereliction which had so marred central Dublin in the 1980s. It also halted decades of inner-city population decline.

6.8.1 Misgivings

Despite their impact, the renewal policies had their critics. Problems included the over-supply of space, the negative effects on surrounding areas located on the 'wrong' side of the incentives line and the problem of competition with existing businesses from those benefiting from subsidy. The sporadic geographical extension of the designated areas also limited the likelihood of the emergence of a local multiplier resulting from greater spatial concentration of incentives. Reliance on individual property owners to initiate schemes also minimized opportunities for co-ordinating development more widely.

Negative consequences also arose for inner-city communities. As the redevelopment potential of land escalated, pressure increased on low-grade businesses to capitalize on the value of their sites. Businesses which had provided blue-collar employment appropriate to local skills disappeared as their sites became redeveloped for offices or enclaves of high-cost housing well beyond the affordability of locals (MacLaran, Williams and Emerson, 1995). It even became difficult for community-based training and local employment schemes to find affordable premises. Moreover, insofar as the problems of the indigenous communities were being diluted by significant numbers of young middle-class professionals moving into the inner city, politicians were now able to point to falling unemployment rates and the rising social profile. Yet little had actually been done to alleviate conditions of those in poverty or experiencing long-term unemployment.

6.8.2 The impact on planning

The impacts of government-sponsored renewal together with the new entrepreneurial agenda within Dublin Corporation had a major impact on local-authority planning, effectively marginalizing it and engendering feelings alienation among staff (McGuirk and MacLaran, 2001). The desire for renewed relevance necessitated a major reorientation of planning towards an entrepreneurial approach facilitative of property capital rather than some nebulous concept of defending the 'public interest' or 'common good'. Planning would clearly have to depart significantly from its historically bureaucratic functions of zoning and development control. Such reconfiguration would need to accommodate a more 'holistic' approach, a mechanism for levering cross-sectoral funding and other resources necessary for the implementation of planning strategies and to adopt an institutional and management structure which could accommodate a broader interpretation of its role. This suggested a role for urban planning concerned less with matters of land-use control and regulation and more with the development and implementation of broad and holistic development strategies. It would involve formulating strategies for small areas, then acting as a catalyst and 'enabling authority', flexibly facilitating and co-ordinating the strategies of public and private sectors alike to mobilize the synergies which might arise from such collaboration (McGuirk and MacLaran, 2001).

This local-area approach became embodied in the idea of Integrated Area Plans (IAPs), a holistic and integrative approach to urban planning which mirrored the planning approach of the special purpose development agencies, such as the Dublin Docklands Development Authority. IAPs are localized planning mechanisms that aim to embrace the complexity of contemporary urban systems through developing a holistic approach towards the achievement of social, economic and environmental goals, while encouraging the necessary inter-sectoral co-ordination to achieve such aims. The Dublin City Development Plan (1999) aims to be a 'more stream-lined strategic plan, capable of responding to the complex development needs of the city' (Dublin Corporation, 1999, 9). IAPs comprise a core element, providing a locally flexible planning framework capable of coping with the complexity and variability underpinning urban development conditions. The approach emphasizes planning's catalytic role and its integrative functions, marking a major repositioning.

However, while planners sought renewed relevance through micro-area planning, the lack of co-ordination between planning authorities and the weaknesses of the urban planning system at the metropolitan scale were generating major difficulties.

6.9 Development in the 1990s: new problems and Edge City

While the inner-city became the focus for renewal under government-sponsored property development initiatives, burgeoning economic growth continued to create a huge demand for buildings, particularly in suburbia. Ireland's GDP grew strongly through the 1990s, at an average annual rate of over 5 per cent, compared with a EU average of 1.5 per cent. With high birth rates lasting well into the 1980s, the youthful age structure contributed to rapid household formation and an escalating demand for dwellings. Adults who might previously have been content to live with parents into their late 20s now sought accommodation of their own, free from moral supervision. Full employment and rising incomes permitted them to fulfil such desires at an earlier age. Rising incomes also created a blossoming consumer market for every conceivable item of consumption, from mobile phones to personal computers, wide-screen televisions to BMWs. Car ownership rates rocketed, as did traffic congestion.

The division of the metropolitan area into four separate jurisdictions had created a fragmented administrative structure with little co-ordination. Deprived of income from domestic rates (property taxes), the suburban local authorities increasingly vied with one another to attract valuable office and retail developments, reminiscent of the 'fiscal mercantilism' found in US cities (see Johnston, 1979). Planners came under pressure to facilitate such income-generating development by introducing new land-use zoning categories, notably that for 'office-based industry', which was often applied to areas previously zoned for industrial functions (Bertz, 2002). Simultaneously, office suburbanization was further stimulated by the declining availablity of inner-city sites and growing competition for them from the buoyant hotel and luxury residential sectors.

The late 1990s also saw a diversification in user demand for office space, related in particular to the influx of foreign companies in the high-technology, computer software and teleservices sector. Many had little need for central-city locations, requiring instead large amounts of high-specification space, good telecommunications infrastructure, capacity for expansion and cheap rents, especially in the cost-conscious teleservices sector (Bertz, 2002). Such criteria were generally met only in suburbia. Prime office rents, unchanged at around Ir£160 per sq. m. (Ir£15 per sq. ft.) from 1990–5, reached Ir£ 400 per sq. m. (Ir£37 per sq. ft.) in 2001. In contrast, space at the periphery was available for as little as Ir£150 per sq. m. (Ir£14 per sq. ft.), while in some locations, tax incentives for occupiers increased their attractiveness and the likelihood of developers' attracting long-term investor financing.

The newly-developed peripheral campus-style office parks also had space for expansion, provided for ease of access by car from the new circumferential M50 motorway and benefited from planning regulations which permitted ample parking. The M50 itself, originally planned as an outer by-pass, became the linear focus for the development of industrial estates, office parks and regional-scale shopping centres, dotted along its length like beads on a necklace. Though still incomplete in its southern section, the motorway is already subject to congestion and delays at its interchanges.

Moreover, rapid growth in car ownership levels, increasing amounts of inter-suburban commuting and worsening traffic congestion created a situation where commuter access to peripheral employment nodes, such as Citywest, is now often quicker from as far afield as Carlow (70 km.) than from inner-suburbs such as Harold's Cross (10 km). The expansion of employment at the urban periphery has therefore resulted in a widening of Dublin's commuting field to 90km. around the metropolis (Williams and Shiels, 2002). Amazingly, little of this appears to have been foreseen by planning authorities.

Nowhere is the development of Edge City better exemplified than in the prestigious southern suburb of Sandyford, some 8 km. (5 miles) south of the city centre. Capitalizing on the presence of prestigious companies such as ICL, Microsoft, Oracle, Trintech and Allied Irish Banks, an industrial estate located near the M50 motorway and soon to be connected to the city centre by a light rail line is being transformed. Benefiting from re-zoning to office-based industry by a local authority keen to increase its income from business rates, substantial office development has occurred. From a stock of 5,967 sq. m. (64,200 sq. ft.) in 1990, Sandyford is developing into a suburban office node with a stock of over 143,250 sq. m. (1.54 million sq. ft.) at the end of 2001. Space has been taken up by Eircell/Vodaphone (the major Irish mobile phones operator), First Active Bank, Bank of Ireland and Barclaycard, with additional space being taken by Microsoft. Development has occurred at green-field sites and also involved redevelopment of low-value industrial space. Thus, single-storey warehouses with a workforce of a dozen people might be replaced by multi-storey office buildings accommodating several hundred employees. With a transportation infrastructure geared to a numerically small employment base, traffic congestion has been the inevitable consequence.

More generally, in view of the lack of co-ordination of land-use and transport planning, the authorization by planners of the proliferation of peripheral employment nodes is disquieting (MacLaran and Killen , 2002). Public transport systems function best when connecting a multiplicity of origins (residences) and a single destination, normally the city centre. However, the evolving pattern of contemporary office suburbanization has created innumerable employment destinations. The outcome is a growing complexity of inter-suburban journey-to-work patterns, reminiscent of American cities, which cannot easily be accommodated by public transport. The high level of enforced dependency on the car is now causing serious traffic congestion in suburbia, inevitably encouraging demands for expensive suburban road improvements.

Only the economic downturn of 2001–2 and the over-provision of space in some peripheral office campus developments have provided some respite in development activity. In one such scheme, vacancy approached 70 per cent in June 2002, quoted rentals fell by over 40 per cent in the course of a year, development finance and long-term funding evaporated and construction was suspended.

This pause in development has permitted time for reflection about the changes which have transformed Dublin during the 1990s. A combination of interventions by the national government, a deepening entrepreneurial local administrative and planning ethos together with an environment of competition between local authorities for commercial property development provided the context for change. But above all, it was the activities of the private-sector property development forces which dominated urban development, and it is to the location and character of those developments that the appropriate transport infrastructure will now have to be adapted. Inevitably, the taxpayer will be obliged to foot the bill, which will be enormous.

7

Remaking the city: property processes, planning and the local entrepreneurial state in Auckland

Laurence Murphy

7.1 Introduction

In contrast to its rural and wilderness image, New Zealand is a highly urbanized society. In 1991, 85 per cent of New Zealanders lived in urban areas and 68 per cent of the population lived in centres with a population in excess of 30,000 (Lees and Berg, 1995). The country's urban system is dominated by the Auckland Region, located in the North Island and having a metropolitan population of 1,074,502, which accounted for 29 per cent of the national population in 2001. As the largest urban centre, the city occupies a strategic position in the national economy. Moreover, within the context of the wholesale restructuring of the economy since the 1980s (Kelsey, 1997), Auckland has assumed an important role as a conduit of globalization processes (Murphy et al., 1999). Consequently, it has experienced significant transformations in its demographic structure, ethnic composition, nature of local governance and, of particular significance to this chapter, its built environment. In the 1980s more office space was built in Auckland's CBD than was built in Paris, a global city. In the early 1990s, Auckland had the second fastest rate of population growth in Australasia (Austin and Whitehead, 1998) with the majority of this growth attributable to international migration, primarily from Asia. During the 1990s, high-rise apartment developments rivalled office towers on the CBD skyline, reflecting the increasing popularity of downtown living within the city. Indeed within the last 30 years Auckland has undergone major cultural and physical transformations.

To appreciate the scale and rapidity of urban change that has occurred within the Auckland region it is necessary to be aware of the dynamics of New Zealand's social and political transformation since the 1980s. Traditionally the state has occupied a dominant position within the New Zealand economy. State agencies dominated infrastructural industries, land-based productive industries and key services. The state accounted for 20 per cent of investment, produced 12 per cent of national income and was the largest employer in the country (Britton et al., 1992). The economy up to 1984 was highly regulated, with significant controls on both the inward and outward flows of capital. New Zealand's welfare state was constructed around policies of full employment, high wages and targeted welfare benefits for those who were in need. During the 1980s, in response to poor economic performance, the fourth Labour government embarked upon a set of neo-liberal economic reforms that ushered in 16 years of considerable economic and social restructuring. The role of government was altered radically through corporatization (the remodelling of government agencies along business lines) and then privatization. Indeed, the scale of New Zealand's privatization programme was comparable to that of the Latin American experience (Pawson, 1996). In effect New Zealand adopted a mode of governance that embraced neo-liberal ideologies of efficiency and market-led reforms, doing so with an explicit aim of engaging with processes of globalization. As a key element of the national economy, Auckland has come to bear the imprint of these seismic reforms in its 'restless urban landscapes'. (Knox, 1993).

This chapter has as its focus the sets of agents and processes at work in the (re)development of Auckland's built environment since the 1980s. The Auckland region consists of four territorial authorities (Auckland City, Manukau, North Shore and Waitakere) and three districts (Franklin, Papakura and Rodney) (see Figure 7.1). The focus of this chapter is Auckland City, which is the authority with the largest population and includes the CBD, which is a major hub of economic activity in the region. Throughout the chapter, I am concerned to highlight the changing dynamics between

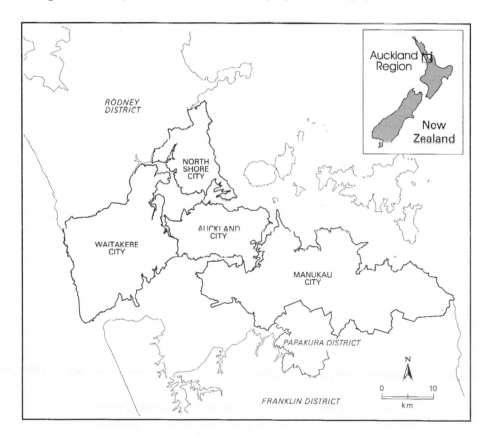

Figure 7.1 Territorial authorities in the Auckland region

agents involved in the creation and reproduction of urban space. The chapter sets out the broad contours of New Zealand's planning system as a prelude to an examination of office development processes, apartment development and public–private urban redevelopment schemes centred on the creation of landscapes of consumption and spectacle. While Auckland's urban development process conforms to trends found elsewhere in the world, the specific character of change reflects the intersection of global and local processes articulated through nationally constituted 'structures of provision' (Ball, 1986, 1994).

7.2 The New Zealand planning context

New Zealand's earliest planning legislation, the Town Planning Act 1926, required cities and boroughs, to prepare towns plans designed to regulate land-use. The principal mechanism for land-use regulation was zoning, in which areas were designated for specific classes of land development. Memon (1991) argues that this formative period of urban planning was characterized by zoning practices that were designed to ensure a degree of certainty for landowners and developers rather than as a vehicle for implementing wider community concerns. Within the context of a colonial society, built upon an ethic of growth and resource exploitation, societal values emphasized 'property ownership, "freedom of the individual" and materialistic attributes' (Memon, 1991, 25). In addition, the practice of planning became increasingly subject to legal interpretation and dispute. Consequently legal emphasis on the rights of the individual and property rights meant that the rights of the public were often relegated to a secondary position.

Reviews of the planning legislation in the 1970s resulted in the recognition of the role of planning in protecting and advancing community values. In particular, the Town and Country Planning Act 1977, required councils to take account of end users and, significantly, the needs of Maori people (the indigenous people of Aotearoa/New Zealand). Under this legislation 'third party' rights were recognized. Thus the right to object to a development was extended not only to the parties affected, but to 'any person representing any relevant aspect of the public interest' (Memon, 1991, 27). Reflective of the legislative changes, the planning process moved from an emphasis on the physical ordering of the built environment towards a greater emphasis on the provision of social, cultural and environmental amenities. Notwithstanding the significance of the 1977 Act, Memon highlights the conservative nature in which this legislation was implemented and interpreted. In general the strong legal and adversarial context in which planning was situated resulted in a context whereby private property rights were protected. The evolution of the planning process clearly legitimated the role of the local state as an agent of land-use planning but the role of the state (both central and local) was, and continues to be, problematic. It is important to note that local authorities constitute the regulatory agency for land-use planning while simultaneously they have significant powers to engage in speculative property development processes (Gunder, 2000).

New Zealand's shift to a neo-liberal, market oriented, economy in the post 1984 period held significant implications for the planning process. Underpinning the wholesale reforms of the economy was a belief that the market was the most efficient mechanism for allocating scarce resources and that the legitimate role of the state was one of minimal intervention. Within the context of the significant reorganization of government, the role of planning was scrutinized and profoundly altered as a consequence of the resource management law reform process which culminated in the repeal of the Town and Country Planning Act and the introduction of the Resource Management Act, 1991. The Resource Management Act (RMA), whilst clearly designed to promote the market allocation of resources, was strongly influenced by environmental concerns and has as its overriding objective the sustainable management of resources.

The RMA has been subject to considerable analysis and critique (Dixon et al., 1997; Gleeson, 1995; Gleeson and Grundy, 1997; Grundy and Gleeson, 1996). With respect to urban property development and planning practice, the Act has had profound impacts. The RMA

> sets up a new style of planning for managing resources which is effects based, representing a shift from the direction and control approach of the past to a focus on environmental outcomes or effects. There is also a shift from the presumption that land use activities are prohibited unless specified to a presumption that activities are allowed unless otherwise specified in the district plan
>
> (Dixon et al., 1997, 605)

The emphasis on environmental outcomes is primarily constructed in bio-physical terms and it is significant to note that the Act makes no use of terms such as 'town, city or even settled area' (Grundy and Gleeson, 1996, 203). More specifically, it has been argued that the new legislation 'eschews any substantial consideration of socio-economic concerns in its purpose' (Gleeson and Grundy, 1997, 299).

In terms of planning practice the RMA has a number of implications. The move to an effects-based regime potentially undermines the role of zoning. Developers can pursue a development as long as they can minimize environmental externalities and/or obtain consent from community interests. Section 94 of the Act sets out the circumstances by which developers can avoid the public notification of their consent application. Non-notification can occur in situations where the council is satisfied that the adverse effects of a development are minor or when

> written approval has been obtained from all those whom the council thinks will be adversely affected, unless it considers it unreasonable in the circumstances to obtain the written approval of every such person.
>
> (Gleeson, 1995, 44)

The non-notification of resource consents is viewed as a mechanism for enhancing efficiency in the planning system in that it can result in a reduction in social costs and time involved in the creation of a development. For planners, the non-notification process provides considerable challenges as they are required to determine the impact area of a proposed development and identify individuals who have property interests within this area. Having identified the various interests, they are then required to assess whether the impacts are minor. These tasks are complex and Gleeson (1995) highlights a number of problems facing planners in their decision-making process. For developers, the system offers the potential to purchase written consents, thereby removing the right from the vendors of those consents to object to a development. Moreover, non-notification offers an effective mechanism for developers to avoid costly public opposition to controversial developments. While Gleeson (1995) has highlighted the problems of the non-notification process, especially in light of the capacity for developers to buy written consents, it should be remembered that non-notification arises primarily when the proposed activity is provided for under the local district plan. In this context, public concern over proposed developments needs to be embedded in the formulation of these plans. Whether citizens are motivated to partake in the development of district plans and can negotiate the technicalities of these substantial documents is a moot point.

Despite the shift to effects-based planning, zoning has continued to be evident in district plans. However, the new system has resulted in a more flexible interpretation of zoning at the local level (Dixon et al., 1997) and has encouraged a 'geographically diverse regulatory regime, which may cause investment uncertainty and thus become a source of frustration to development capital' (Gleeson and Grundy, 1997, 301). Section 94 seems to have resulted in a significant reduction in the proportion of consents that are publicly notified. Grundy and Gleeson (1996, 209) found that 'Auckland City Council now only notifies 8–9 per cent of consent applications, as against 22 per cent before the introduction of the new Act'.

In terms of property development processes, the shift to an effects-based planning system reinforces the primacy of the market as the key regulator of new development. In practice, the operation of the RMA has engendered considerable debate regarding the extent to which the system is subject to delays, court hearings and time-consuming bureaucracy. Clearly, the RMA adds to a long history of planning practice that is supportive of private property rights by creating the conditions for greater public–private interaction in the development process. In common with trends found elsewhere in the world (see Dear, 2000; Fainstein, 2000) planning in New Zealand has become more flexible, facilitative and entrepreneurial.

7.3 Office development processes: agents of change[1]

Auckland City has experienced considerable office development activity since 1975. The CBD has traditionally been the dominant locus of new office developments although in recent years suburban office parks have increased in popularity. Figure 7.2 shows the amount of new office space started in the CBD each year for the period 1975 to 2001. In total, more than 967,000 sq. m. (10.4 million sq. ft.) of new office space was built during

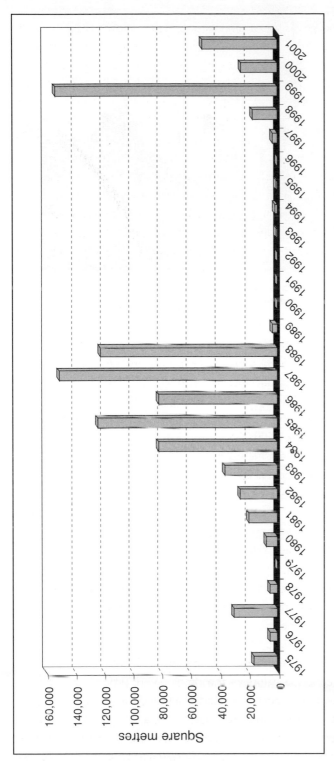

Figure 7.2 *Office development new starts 1975–2001*

Figure 7.3 *Auckland CBD with Sky Tower in the centre*

this period, with development occurring in distinct development cycles. Four phases of output can be identified. From 1975–83, a total of 147,471 sq. m. (1.6 million sq. ft.) was developed with an average annual output of 16,385 sq. m. (176,000 sq. ft.). Between 1984 and 1989, output increased dramatically with the average annual output of new space increasing to 94,250 sq. m. (1 million sq. ft.) and the total amount of new office space in the CBD increasing by more than 500,000 sq. m. (5.4 million sq. ft.). This scale of development was unprecedented for New Zealand (Moricz and Murphy, 1997). The aftermath of the 1987 global sharemarket crash had a significant impact on office demand and the financial position of property developers, resulting in a dramatic slump in office starts between 1990 and 1996. Only 3,671 sq. m. (39,500 sq. ft.) of new space was built in the CBD in this seven-year period. Since 1997, a new round of development has occurred in the city with just over 250,000 sq. m. (c.2.7 million sq. ft.) of office space being developed.

The boom and slump nature of office development in Auckland's CBD seems to conform to traditional demand-led explanations of property development, especially when one considers that during the mid-1980s boom, vacancy rates were below 5 per cent and rents jumped from NZ$70 to NZ$300 per sq. m. (NZ$6.50 to NZ$27.90 per sq. ft.)(Moricz and Murphy, 1997). Yet, throughout the different cycles of development significant changes occurred in the nature of the development process, the agents of supply and the manner in which property operated as a financial asset.

In the late 1970s, financial institutions such as banks and insurance companies dominated the development process. However, over time, property companies became the key agents in the commercial property market. Between 1975 and 1990, property companies were the single most important initiators of new developments accounting for 45 per cent of the space developed (323,699 sq. m. (3.48 million sq. ft.)). Significantly, one corporation, Chase property group, accounted for 53 per cent of the output of all property companies and 24 per cent of all new developments in the CBD. The emergence of these property companies was underpinned by transformations occurring in the sharemarket.

The sharemarket, and individual property company share price performance, provided vital ingredients in the development of property developers' strategies. A persistent bull market throughout the mid-1980s meant that substantial funding could be obtained by direct share placements. Institutional investors, both local and international, were attracted to New Zealand equities as share prices trebled between 1984 and 1986. Significantly, 'good market' signals meant that property companies were strategically placed to attract large-scale funding from the recently deregulated domestic banking sector. In the context of a booming commercial property market, developers arranged funding on the basis of individual schemes and, in retrospect, it seems that the banks were less concerned with the overall financial character of developers and more concerned with accessing the benefits of the boom. In time, the strong performance of property companies' shares provided the means for these companies to access international capital markets. For example, in 1985 Chase Corporation Ltd. became the first listed company in New Zealand to issue unsecured Swiss Convertible Notes and, a year later, it issued unsecured Swiss bonds. Both issues offered European investors the opportunity of subscribing to Chase shares.

In contrast to traditional investment strategies that construct property as a long-term investment, funding derived from the sharemarket places greater emphasis on short-term performance. Changes in accounting practices in the 1980s allowed companies to revalue properties on an annual basis and to include any increase (or decrease) in values in their statement of profits (Newland, 1994). Rising rentals, especially prime CBD rentals, ensured the growth of capital values on all new developments and allowed property companies to occupy a favoured position in the sharemarket. The increasing importance of the sharemarket as a source of funds had implications for the manner in which property agents operated. As Pryke (1994b, 239) puts it:

> ... the capital value of company shares will be determined heavily by its ability to sell developments at or above the costs of production and/or to retain high-yielding schemes in its investment portfolio. Institutions holding property company shares ... will require satisfactory income returns and capital growth. Market performance must ultimately be met.

An examination of the activities of the major agents in property development during this period highlights the different ways in which property was used as a vehicle for profit generation and asset growth. Construction companies, whilst major developers of new space, retained ownership of only 2.4 per cent of this space. Reflecting their underlying rationale as construction firms, these companies were concerned with realizing development profits. In contrast, property companies remained owners of 32 per cent of the new space developed in the CBD. Whilst development profits provided useful inflows to these companies, the retention of developments allowed them to record significant asset growth, thus satisfying the appetite of the sharemarket for successful growth. Chase Corporation's development strategy highlights the importance of developing and investing in the right areas. Chase developed four times as much space in the CBD (172,969 sq. m. (1.86 million sq. ft.)) compared to peripheral areas beyond the city centre (42,211 sq. m. (454,000 sq. ft.)). Moreover, Chase retained 57 per cent of the space that it developed in the CBD compared to only 36 per cent of its non-CBD developments (Moricz, 1994).

By 1986, stockmarket analysts noted that the public perception of the sharemarket was linked to the 'glamour associated with property stocks' (Buttle Wilson, 1987, 10). Robert Jones Investment Ltd, one of the largest property companies in the country, recorded a 479 per cent increase in its share price between January and November 1986. Property company directors became media personalities and media attention was increasingly focused on their property deals.

The history of Chase Corporation highlights the significant interconnections between sharemarket performance and property activity at this time. Established in 1970, as a small property company, Chase enjoyed a dominant position within the New Zealand property market throughout the 1980s. Between 1982 and 1987 Chase's shareholders' fund rose from less than NZ$2 million to NZ$139 million (Moricz and Murphy, 1997). On the eve of the global sharemarket crash, Chase announced an unaudited net profit of NZ$105.3 million, with 85 per cent of profits coming from its core property interests in New Zealand (Chase Corporation Ltd., 1987). Reflecting the phenomenal growth of the company, Chase had developed property interests in Australia, the UK, the US and Hong Kong and had diversified its operations through wholly-owned subsidiaries such as Farmers Trading Company (a New Zealand department store) and Hanimex Corporation (Chase Corporation Ltd, 1987). In 1987 Chase's assets rose sixfold to NZ$3.6 billion. The funding required to sustain this growth included a NZ$461 million share issue and, more significantly, an increase of NZ$1,696 million of debt, raising the overall indebtedness of the company to NZ$2 billion (Moricz and Murphy, 1997). Chase managed to incur this level of borrowing by employing complex accounting procedures that concealed the total level of its debt and by organizing finance on an individual project basis. The overall performance of the company was enhanced by Chase's practice of including unrealized development gains as profit. Chase's property investments thus provided the basis for the expansion of a vast amount of fictitious capital.

The sharemarket crash in 1987 impacted heavily upon property companies and developers who at the time accounted for over 9 per cent of total sharemarket capitalization. Despite continued output, the problems of the sector were becoming manifest. In 1988, Chase recorded a loss of NZ$31.7 million 'but then classified

NZ$109.5 million of the losses as extraordinary, to give an above-the-line profit of NZ$77.8 million' (National Business Review, 1994, 34). Chase, the largest developer investor, was struggling to maintain its operations and service its debt. Its 1989 Annual Report showed an operating loss of NZ$205 million and abnormal and extraordinary losses of NZ$841 million. The bulk of this loss was attributable to the

> write-down of assets in the Group Balance Sheet to their estimated current Net Realisable values. These values most particularly reflect the drop in value of Australian and New Zealand property assets. . . . In some cases Net Realisable values are recorded as being lower than the level of debt secured against those assets.
>
> (Chase Corporation, 1989, 4)

By September 1989, Chase was in receivership. The sharemarket crash impacted on the property sector by depriving property companies of ready access to funds and by undermining user demand. Throughout the property boom, vacancy levels had remained below 5 per cent, but from 1987 vacancy levels soared to above 25 per cent in the CBD. The significant rise in vacancies depressed prevailing rental levels and undermined the prospects of future rental increases. Consequently, the capital value of existing developments slumped and the viability of future developments diminished, especially since prime yields rose in the face of declining institutional interest in property. Thus, property companies were exposed to a dramatic slump in funding at the same time as capital values slumped.

Depressed rentals and sharemarket prices, in conjunction with massive debts, resulted in a virtual collapse of the property investor sector. Faced with the prospect of considerable losses in the property sector, the banks moved to secure the repayment of loans. Property companies, such as Chase, were forced to embark upon radical restructuring programmes aimed at generating cash inflows. For the most part, these strategies proved futile. Between 1987 and 1989, the total market capitalization of property companies on the New Zealand stock exchange declined by 78 per cent from NZ$5,782 million to NZ$1,278 million. Of the top 20 property companies operating in 1987, 11 were in receivership, defunct or subject to takeover by 1989. The demise of these companies was as dramatic as their advent.

The massive output of new space in the CBD during the late 1980s created the conditions for significant oversupply and this situation persisted until the mid-1990s. Vacancy rates peaked at 42 per cent in 1991 and there was thus little incentive for new development to occur. A new round of development began in 1997. This development cycle was triggered by demand for new premium-grade office space, with large floor space and the potential for more flexible space utilization for tenants. In 1999, the 39-storey Royal Sun Alliance Tower, the largest office tower in New Zealand was begun. The internal layout of this tower allowed Royal Sun Alliance to reduce its space requirements per employee from the city norm of 21.5 sq. m. (230 sq. ft.) to 13 sq. m. (140 sq. ft.). Significantly, this new round of development has been undertaken by property companies backed by institutional investors and development has occurred only after significant pre-letting has been secured. This latest round of development is clearly set within an institutional context that does not wish to reproduce the speculative bubble of the 1980s.

7.4 The inner-city apartment market

In the context of high vacancy rates and a significant property slump, secondary office space became redundant and uncompetitive. Faced with empty office space, owners and developers switched their attention to apartment development. In the early 1990s, the first conversions of office and warehousing space into residential apartments began. Notwithstanding the paucity of services for the small existing CBD population, these conversions proved popular. Between 1991 and 1996, a total of 2,015 apartment units were completed in the inner-city/CBD area (Connor, 1997) and the population of the CBD (Auckland City CAU) doubled from 1,404 to 2,835. Since 1996, the scale of individual apartment developments has increased as the market shifted from office conversions to purpose-built apartment blocks. In 1999, it was estimated that by the end

Figure 7.4 Inner-city apartments, Auckland

of 2000 the stock of apartments in the inner-city would exceed 5,500 units (Bayleys Property Research, 1999). As a result of the apartment boom, the resident population of the CBD rose to 8,500 in 2001.

It is possible to discern several trends within this central-city apartment boom. First, the initial focus of development was on office conversions. In effect, the office crash provided a context for the occurrence of what Harvey (1982) has termed 'creative destruction'. In this context, the destruction of capital values in the office sector provided the opportunity for conversion to new uses. Second, the demand for these properties came from owner-occupiers. These included young professionals and older retired couples (empty nesters). Third, once the market was established, there was a distinct shift toward the development of purpose-built apartment blocks. These new developments included serviced apartments and were designed to attract property investors. These serviced apartments are developments in which apartment owners can have their properties managed by a hotel/management group that guarantees a set return on the property. In effect, the serviced apartment market can be viewed as an extension of the hotel sector in the city. Fourth, as the market has expanded, it has become increasingly differentiated on the basis of quality and location. Early office conversions now have to compete with purpose-built apartment blocks that include swimming pools, gyms and restaurants.

Using unpublished official property data, Figure 7.5 depicts the number of inner-city apartment sales on an annual basis for the period 1990 to 2000. From a low start in 1990

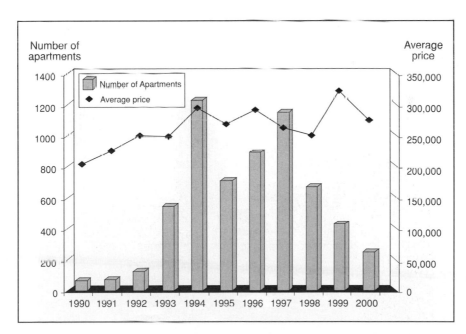

Figure 7.5 Auckland CBD apartment sales (number) and average sale price, 1990–2000

(65 sales), sales activity increased rapidly in 1993 and peaked at 1,228 sales in 1994. They dropped to 700 in 1995 and rose again to 1,150 in 1997, only to decline thereafter. While some caution should be exercised when viewing these figures, as there may be problems in the number of sales recorded (especially for the year 2000), it is clear that this market is subject to considerable annual fluctuation. Indeed, the successful defence of the America's Cup in 2000 promoted a new round of apartment development in the CBD. With respect to sale prices, average prices rose from NZ$197,000 in 1990 to NZ$286,000 in 1994. This early growth in average prices probably reflects changes in the quality of apartments being sold, rather than rapid capital appreciation. Post-1994 average sale prices for apartments have remained within a band ranging from NZ$250,000 to NZ$280,000. The increase in average prices in 1999 reflects the sale of a small number of very expensive apartments.

An analysis of the apartment sector provides insights into the manner in which global flows of people, finance and ideas are embedded in the local built environment (see Clark and Lund, 2000). The aptly named Metropolis apartment building offers an interesting case study for exploring these emerging global-local connections. Completed in December 1999, this 38-storey, 375-apartment building constitutes both a material and symbolic conduit of globalization processes. Developed by Andrew Krukziener, this NZ$180 million development represents a triumph of personal ambition and creative financing. Metropolis is at once the product of a set of very specific local property processes embedded in a wider set of transnational relations. In order to unpack this web of relations I will focus on three elements of the development: its financing, its design and the management of the serviced units.

Figure 7.6 *The Metropolis apartments development (centre)*

Given the ambitious scale of the development, the financing of Metropolis was always likely to be complex. The initial first mortgage on the building was derived from a consortium of international banks including HSBC, while a second mortgage was secured from Deustche Bank. Soon after completion the development was refinanced with ANZ bank obtaining a mortgage on the property. In addition, the developer organized a NZ$21 million junk bond issue, promising a return of 14 per cent to investors. These bonds were a form of subordinate debt and repayment depended upon sufficient apartment sales to cover the mortgage to ANZ, with any excess going to cover the capital and interest due to bondholders. In May 2001, Krukziener defaulted on the bonds and a year later was still involved in negotiations to refinance the deal. Clearly, Metropolis represents a physical embedding of a set of international and local capital flows.

Symbolically, Metropolis encapsulates a set of global metropolitan processes. The very name of the building is designed to speak to different constituent audiences. To the populace at large, it is meant to signify Auckland's big city (global city) status. Its sheer scale is all about putting Auckland on 'the map', in international terms. To investors and/or apartment owners the name Metropolis alludes to the self-contained nature of the building. A city within the city, the building contains a large gymnasium, a 22-metre pool, several exclusive restaurants and retail space. In addition to its name, the building's design and outfitting draws together a set of global ingredients. Although strongly influenced by Krukziener, the design was produced by New York architect Paul Katz, in collaboration with a local design team. The entrance foyer makes use of specially imported Italian marble chosen by Krukziener and art deco themes are evident throughout the building. In terms of ideas and materials, Metropolis is a locus of international flows.

The management of the serviced apartments within Metropolis represents another set of transnational connections. On completion, 315 units were made available for investors as serviced apartments. The initial management of these apartments was granted to Somerset Holdings Ltd. This Singaporean company formed part of the property group Pidemco which in turn was owned by the Singapore government. Through this management relationship Metropolis was incorporated within a chain of 'upmarket' hotel operations. In November 2000, Somerset Holdings merged with The Ascot Group, to form what they term, 'Asia-Pacific's largest serviced residence company' (Ascot, 2001). This newly formed entity owns or manages over 6,000 serviced residential units in 16 cities in 10 countries.

As a case study, Metropolis offers a window into a set of processes. In terms of its financing, design and management, Metropolis exemplifies the manner in which property markets are being globalized. Moreover, the trend toward downtown apartment living accords with what authors such as Ley (1996) and Smith (1996) term the remaking of the inner-city. In effect, the consumption needs of the new middle classes, it is argued, are shaping the character of the built environment in inner-city areas. Significantly, the consumption needs of this group are centred on what has been termed symbolic capital, where the item consumed attests to the cultural discernment of the consumer. The importance of symbolic capital is clearly evident in the marketing of Metropolis and other apartment developments in the city. But at a wider level, cities around the world have increasingly sought to use culture and the built environment as markers of distinction for attracting investment and tourism (Zukin, 1995). In conjunction with the boom in apartment development in the inner-city, local government has itself been actively involved in the redevelopment of the CBD to make it a landscape of spectacle and consumption.

7.5 The entrepreneurial local state and the city of spectacle

New Zealand's engagement with processes of neo-liberal reform has had significant implications for the state and its role in society. The discourse of 'core business', for

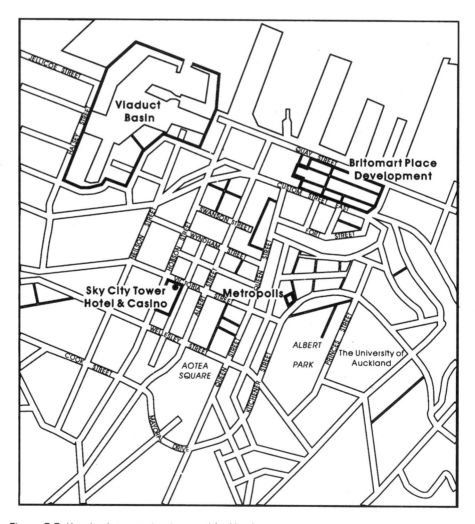

Figure 7.7 *Key development sites in central Auckland*

example, has been used to legitimate or abandon specific functions of central government and the relationship between the local and central state has been fundamentally altered. Within the context of profound economic restructuring the local state has increasingly adopted entrepreneurial forms of governance, centred on place promotion (see Pawson, 1999) and, in Auckland's case, property development. Since the late 1980s, Auckland City Council has sought to strengthen the central city area's role as a key locus of business activity and as a centre of tourism and entertainment. In the 1980s, the council created a business unit to promote business activities and cultural festivals. The council also retained the services of Saatchi and Saatchi to create a city logo. But this proved to be a failure. More significantly, the council has shown a distinct willingness to engage in property-led urban regeneration processes. Three projects in which the council has been involved highlight the entrepreneurial nature of their activities. These are: the Sky City casino development, the Britomart Transport Centre and the Viaduct Harbour (America's Cup Village). These sites are located in Figure 7.7.

7.5.1 Sky City casino

The Casino Control Act (1990) allowed for the 'development of licensed casinos in a manner consistent with the promotion of tourism, employment and economic development'. The Act initially provided licences for two casino premises; one for Auckland and one for Christchurch. After an intensely competitive bidding process, the Auckland licence was awarded in January 1994 to Sky Tower Casino Ltd (which became Sky City Ltd). This scheme was 80 per cent owned by the New Zealand transnational investment company Brierley Investments Limited (BIL) with the remaining 20 per cent taken up by the US based company Promus, owners of Harrah's Casino Hotels. The success of the Sky Tower bid reflected the ambitious scale of the NZ$320 million project which included a casino, conference centre, bars, shops, a multi-storey underground car park and a 344-bedroom hotel (Murphy, 1996). The bid also included a 328 m. Sky Tower with observation decks, restaurant and telecommunications facilities. The Sky Tower, completed after the casino was opened, is the highest tower in the southern hemisphere. In their bid, the operators of the Sky City Casino argued that their facility would attract 3.5 million visitors annually, with approximately 500,000 of these being from overseas. In addition, they estimated that 1,650 construction jobs would be created during construction and a further 4,946 jobs (both on and off site) would be created once the project was completed (Murphy, 1996).

The ambitious nature of this bid conformed to the spirit of the legislation and was particularly attractive to a council keen to see the CBD as a centre for entertainment and tourism. However, the potential viability of this scheme was jeopardized as a consequence of the poor location of the proposed casino site. Indeed, the developers failed to secure planning permission for their original project. With the potential collapse of this bid, Auckland City Council stepped in and entered into a land swap arrangement with the developers. The city council swapped a large downtown site (see Figure 7.7), assembled for the development of a public transport centre, for the Sky City Casino site on the edge of the CBD. This land deal considerably enhanced the value of the Sky City bid and ensured that planning permission would be granted for the proposal. In effect, the council entered into a partnership with the casino developers as a means of ensuring the success of their bid. The council believed that local government support for this project was appropriate given the scale and potential economic benefits that would accrue from the venture.

Since its completion, Sky City has proved to be a very successful commercial operation attracting considerable numbers of visitors. In 1997, Harrah's were bought out and the operation is now wholly New Zealand owned. In 2001, Sky City won the New Zealand Tourist Attraction of the Year award, received over 5 million visits and had revenues of NZ$318 million. Sky City purchased a casino in Adelaide (Australia) and has ownership and management interests in other casinos in New Zealand. It is notable that the initial rhetoric regarding the 'high roller' gambling phenomenon has been effaced by a less glamorous reality. Sky City has consistently expanded the number of low denomination (2-cent) poker machines and introduced a 'loyalty card' that was given to 18,000 regulars.

The commercial success of Sky City would seem to vindicate the city council's decision to support this operation. More significantly, it highlights the council's willingness to support what McCarthy (2002) terms 'entertainment based regeneration'. Such strategies, whilst popular with city authorities are designed to promote very localized places and may have limited growth effects for the city as a whole.

Figure 7.8 *The Sky City complex, with base of Sky Tower to left*

7.5.2 *The Britomart development*

If the casino land deal raised some public concern, the council's Britomart Project proved very controversial. The project's origins lay in the council's desire to develop an integrated transport centre in the central city adjacent to the waterfront. From modest beginnings, the project grew to a NZ$1.5 billion property project involving an underground transport facility with 2,900 car-parking spaces and an above-ground development of 11 high-rise buildings including offices, apartments and hotels. While promoted as a 'no risk' development, the project depended on a complex set of financial relationships between the council and the developer. As the largest single development undertaken by a council, the project attracted considerable public attention, especially the extent to which the council was exposed to financial risks. Key elements of the agreement are set out in Figure 7.9. The most contentious issue related to the above-ground commercial development. Under the scheme the developer was to construct the transport centre for NZ$100m and then construct an above-ground commercial high-rise complex that could be sold for development profit. The deal meant that the city would obtain a transport centre and large-scale waterfront development, while the developer had the potential to realize significant development profit. In order to reduce the developer's risk, the council provided a 'standby take-out facility' covering unsold development sites. Under this agreement, in the event that the developer failed to sell any of the sites, the council would be committed to granting the developer a maximum of NZ$230 million in loans. In return for these loans the council would subsequently have 'first charge over the unsold sites and be entitled to the net sale proceeds of those sites to the extent of

---

The main features of the project agreement are that the council:

- Acquires a transport centre, a traffic underpass (Quay Street East) and public areas at a fixed cost of NZ$100 million, with the developer being responsible for the completion of these facilities.

- Contributes NZ$23 million to heritage protection and infrastructure services and $2.3 million towards associated works.

- Sells Britomart properties to the developer for NZ$56 million, with the sale proceeds providing a loan to the developer for the same amount.

- Provides a 'standby takeout facility' for unsold development sites. The maximum possible exposure is NZ$230 million on the basis that unsold sites are held by the council as a security.

- Obtains resource consents for the project, and forgoes development levies and rates for up to five years.

(*Source:* The Controller and Auditor-General)

Figure 7.9 *Britomart project details*

the full amount then owing to the Council' (The Controller and Auditor General, 1999, 18). In effect, the council guaranteed the developer's development profit and did so in the belief that any loans made to the developer would be repaid by the sale of the development.

The viability of the project came into question as the council had difficulties in securing the planning consents in the Environmental Court. In addition, the chosen developer Savoy Equities, described by one commentator as having 'meagre financial resources' and 'no experience in large-scale property development' (Gaynor, 1999), struggled to secure the appropriate funding. Public concern regarding the project was such that the Office of the Controller and Auditor General reviewed the project in 1999. The Britomart Project proved to be an emotive election issue and resulted in the election of a new mayor and council in 1998, pledged to halt the Britomart Project. After involved legal consultation, the new council terminated its contract with Savoy Equities in November 1999 and opted for a less ambitious project stripped of the high-rise developments. The new 'Britomart' development still involves considerable council funding, in excess of NZ$133 million, but is designed primarily as a transport centre.

The council's involvement in the Britomart Project highlights the extent to which the council was prepared to engage in speculative development processes. It is worth noting that whilst the council believed that it was a 'no-risk' project, the Britomart office developments were being proposed at a time when new office developments were not even being considered by developers unless they had achieved considerable pre-letting. Significantly, the impetus for the project came, not from the council's planning department, but from the Council's Property Development Manager, Peter Cross. Interestingly, Cross was a former chief executive of McDow Properties, a company that had been actively engaged in the production of office space during the 1980s boom.

7.5.3 Viaduct Harbour

Auckland's waterfront has long been the site of contested visions of the harbour. Auckland is the largest port in New Zealand and the activities of the port restrict public access to the waterfront. As waterfront redevelopment schemes have flourished around the world, Auckland City Council has sought to recreate the city's 'natural asset' with an eye to tourism and spectacle. Team New Zealand's victory in the America's Cup in May 1995 provided the impetus for turning latent plans into reality.

While the euphoria of victory in the America's Cup galvanized local political and entrepreneurial will to seek to capitalize on the profitable dimensions of the event, no private interests showed any willingness to develop the Cup base around which these opportunities for profit would exist. The possible sites for development were on the commercial edge of the city and were deemed to be too risky. Team New Zealand and the Ports of Auckland entered into discussions concerning a possible site but were unable to secure funding. In November 1995, Team New Zealand, the Royal New Zealand Yacht Squadron and the Port Company requested the then Auckland Regional

Services Trust (ARST) to take responsibility for developing and financing the facilities for defending the America's Cup. The ARST agreed, on the condition that the investment had a viable financial justification and would be of clear benefit to the people of the Auckland region.

The ARST entered into a consultation process with interested parties and effectively undertook to pay for the cost of developing the bases. The ARST designed the Cup Village (known as the American Express New Zealand Cup Village, the American Express company providing limited sponsorship for the naming rights) and handed over the management of the site to the America's Cup Village Ltd. Various estimates as to the level of public funding involved in the development indicated that Auckland City Council spent NZ$40 million (on public squares, walkways and sea walls) and that the ARST and the central government spent NZ$70 million and NZ$10 million respectively on reclamation, wharves and syndicate bases for the America's Cup. This significant public investment, primarily designed to support New Zealand's America's Cup defence and to showcase the city to the world, substantially enhanced returns to private investors in the Viaduct area. In particular, the infrastructural redevelopment of the area benefited the landowners, Viaduct Harbour Holdings. In a somewhat unusual arrangement by New Zealand standards, Viaduct Harbour Holdings has retained ownership and leased the land on the basis of 20-year perpetually renewable leases. Despite the leasehold land structure, private developers were drawn to the site as a result of the upgraded public facilities and the prospect of significant demand for apartments overlooking the waterfront and close to the action of the America's Cup defence. Owner-occupied and serviced apartments have been developed in the Viaduct, including Latitude 37,

Figure 7.10 *Viaduct Harbour, apartments and restaurants*

Watermark (managed by the Australian company Mirvac, developers of Sydney Olympic Village) and The Point. The successful defence of the America's Cup in 2000 and, in particular, the success of the Viaduct Harbour as a site of entertainment and festivity, has ensured continued investment in the area.

The Viaduct Harbour's property boom has resulted in a significant shift in property development westwards from the CBD and towards the harbour. Given that the council has long cherished the desire to transform the harbour edge into a leisure and events centre for the city, it would seem that the local and central government pump-priming of the property market has been effective. Yet, the emerging landscape of affluent apartment developments and upmarket restaurants suggests that the remodelled habourside has been fashioned in the image of social exclusivity associated with international yacht racing.

7.6 Conclusions

Auckland City and its CBD in particular, have been subject to successive bouts of speculative property development. In pursuit of profit, speculative booms have been followed by significant property crashes, most obviously demonstrated in the experience of the office development market during the 1980s. Clearly, Auckland's built form is at once a testimony to the excesses of the past and the hopes of the future. In this context it would seem that David Harvey is correct when he argues that capital is the architect of the built environment. However, as this chapter highlights, the interests of capital are highly differentiated. The development interests that coalesce around specific projects are spatially and temporally specific. The agents involved in the creation of office space in the 1980s differ significantly from those involved in property investment in the late 1990s. This reflects not only the operation of user demand, but also the manner in which agents and financial institutions construct property as a financial asset. Similarly, Auckland's inner-city apartment boom reflects the operations of both an owner-occupier market and an investor market for serviced apartments. The manner in which profit is extracted from developments (development gain, rental income, capital gain, etc.) and the ways in which property is used as a financial asset (for long or short-term returns) holds significant import for the geography of development.

If the built environment is the 'output' of the property development industry, it is also a semiotic domain, a landscape read by both locals and tourists. Within the context of increasing entrepreneurial governance, Auckland City Council has promoted, or assisted, developments designed to place Auckland on the global map of tourism, culture and consumption. The land deal that secured the Sky City casino development, resulted in the construction of a highly successful tourist attraction with a landmark tower. The Sky Tower has, for its part, become an iconic symbol of Auckland's brashness (Pawson, 1999) and is a marker of its claim for 'world city' status. The ambitious Britomart project demonstrated a significant willingness on the part of the local state to underwrite the risk of development for private developers in return for a landmark development that was believed to have the potential to revitalize the waterfront. The local state's involvement in the Viaduct Harbour represents both an investment designed to capitalize on the media exposure associated with the America's Cup and an investment in public access to the waterfront. Yet, the exclusive nature of residential and retail development around the basin raises tensions between private residents' interests and public amenities. Beyond these issues, the Viaduct is a site for the staging of spectacle and the attraction of wealthy visitors, especially the wealthy owners of super-yachts. Like all developments, it has the potential for devaluation and so it is a place whose value needs to be maintained or policed. Interestingly, its success and future may be aligned with the fortunes of Team New Zealand and not the inherent attributes of the place itself.

This chapter has examined the evolving coalitions of agents and institutions involved in the creation of Auckland's built environment. The city, and especially the CBD, has been the locus of considerable speculative development. The combination of a planning context that is effects-based and permissive toward development and a local government actively engaged in property development, offers the potential for significant rounds of new investment. The success of these property developments should not be measured solely in financial terms as these new spaces are inherently social and have the power to exclude or engage the public. In Auckland, there has been a significant shift in inner-city property development towards the creation of consumption landscapes centred on notions of spectacle and lifestyle. These landscapes are inherently speculative in nature and are subject to changing consumer trends and fashions. Clearly, the developments that have occurred have contributed to place-specific regeneration within the inner-city but the long-term social and economic implications of these developments remain to be charted.

7.7 Notes

1   This section on office development draws substantially from the article by Moricz and Murphy, 'Space traders: Reregulation, property companies and Auckland's office market, 1975–1994' published in the *International Journal of Urban and Regional Research*. The material is used by permission of the copyright holders, International Journal of Urban and Regional Research and Blackwell Publishing.

8

Reshaping and reinventing the city of Birmingham, UK: planning for enterprise and the information society

John R. Bryson

8.1 Introduction

A new regionalism has been identified in the geographical literature (Storper, 1998) based around the identification of new industrial and service spaces (Bryson, 1997a) whose competitiveness is constructed around local knowledge and dense concentrations of human labour. Such places, according to Scott (2002,1), 'constitute distinctive subnational (i.e. regional) social formations whose local character and dynamics are undergoing major transformations due to the impacts of globalization' and, more importantly, 'are foci of significant new experiments in local political mobilization and reorganization'. The difference between this new form of regionalism and older forms is that these localities are increasingly developing their own economic, social and cultural identities and have begun to engage with the global economic system without being completely subservient to the dictates of the central state.

City-regions are distinctive places that engage with national state policy by a process of interpretation, selection and manipulation. This process of engagement is an attempt to shape the relationship between the processes driving economic restructuring and the provision of an environment that will retain and attract investment, as is also evident from several of the other case studies in this volume. The development of a global economy has undermined local advantages that are founded on physical attributes (location, raw materials, etc.) and employment characteristics. For cities located in developed market economies, there will always be alternative locations that are able to undercut production costs due to the availability of cheap labour and limited employment and environmental regulations. The competitive position of a city is now founded on urban cultures, knowledge and expertise and it is these softer elements and the physical structures that support and encourage them which need to be created, enhanced and protected.

In the UK, inter-urban competition for footloose private investment, central government resources and cultural institutions or sporting events has been an important feature of urban policy. From the late 1980s, British cities have engaged in proactive city marketing strategies and environmental improvement campaigns to attract investment. This represents a radical break with traditional planning that took a largely passive and responsive role to urban regeneration. A central component of the new proactive entrepreneurial approach to planning has been the creation of publicly-funded or subsidized flagship developments designed to improve the local environment, attract private-sector investment and enhance the city's reputation as a place to live, work and play.

Exploring the ways in which city-regions control and manipulate nationally formulated urban planning legislation and procedures to achieve local spatial and/or social outcomes is central to understanding the development of the new regionalism. It is an exercise in unravelling the ways in which city-regions are coping with managing the interface between local, national and global processes. This chapter provides an account of this interface by exploring the relationship between property development and planning in the city of Birmingham. Located in the West Midlands of England, Birmingham is the UK's second city with a population of just over one million persons, comprising the largest local authority in the country. The city is part of a much larger conurbation that includes the cities of Coventry and Wolverhampton. Over 5.2 million people live within 80 km. (50 miles) of Birmingham city centre. Birmingham itself is a multicultural city with 21.5 per cent of the population categorizing themselves as other than 'white' in the 1991 census, with the largest single group comprising Pakistanis who accounted for 6.9 per cent of the city's population (Slater, 1996).

The argument is divided into three parts. The first part explores some dimensions of national planning policy in the UK. The second part examines the economy of Birmingham and identifies the pressures that are driving change in the city. The third section identifies three stories, or more correctly 'moments', illustrating the ways in which Birmingham's local planning framework manipulated and has been manipulated by various forms of property capital.

8.2 The national planning context

It is unnecessary here to provide a detailed historical account of the evolution of the UK's national planning framework (see Cherry, 1994; Ward, 2002). All that needs to be explored here is those elements of the national context that have influenced the interaction between Birmingham's local planning context and private-sector development interests. It is important to remember that the national planning framework is interpreted and negotiated as it is applied locally and, in many respects, understanding this process of application is more important than detailed knowledge of national policy. This is a two-way process: the local articulation of national planning policy in Birmingham has influenced national policy by encouraging the state to emphasize urban design frameworks and mixed-use developments as central components of regeneration schemes (Office of the Deputy Prime Minister, 2000).

Two strategies have informed British planning: urban development corporations/ enterprise zones and public–private partnerships. Both strategies reflect a shift in British policy towards one in which 'the engine of enterprise is now the driving force of inner city policies' (Deakin and Edwards, 1993, 1). The first strategy was informed by the political philosophy of Margaret Thatcher's Conservative government. The emphasis was on the state creating the conditions for private-sector investment and, in its most radical form, transferring planning control from the local state to public–private partnerships controlled by private-sector interests. The emphasis was on individualistic, market criteria dominating decision-making as opposed to collective social or community interests (Thornley, 1988). Public money would improve the quality of the environment by improving a designated area's infrastructure (roads and services), by assembling sites that were large enough for profitable redevelopment and by dealing with contaminated land. Policies were also formulated that subsidized uneconomic development propositions by closing the financial gap between total development cost and current rental values (Bryson, 1997b).

Urban Development Corporations (UDC) were established to deal with problem areas, the first two, Merseyside and London Docklands, being created in 1981. UDCs were single-purpose short-life authorities appointed and financed by government and outside the control of local government. They were unelected bodies whose primary purpose was to bring about industrial or commercial regeneration by using public funds to leverage private investment. UDCs had substantial powers to acquire, reclaim, develop and dispose of land and buildings. They formulated planning policy and determined what should be developed and where it should be built. They were also excluded from the planning appeals system and could grant planning permission without consultation or planning inquiries (Brownill, 1990). In planning terms, UDCs removed restrictions and controls placed on development by the local state and transferred control to an unelected board composed of politicians and business interests, including representatives of local and national property development and investment companies.

The second strategy, partnerships, was part of the UDC movement, but it has become a central component of the British political system. Even under the 1997 Labour government, public–private partnerships have played an increasingly important role in designing and developing buildings to house essential public services such as hospitals, schools and prisons.

These two strategies inform the relationship between property development and planning in Birmingham as the city council attempts to manage the development process at a variety of different scales. For large regeneration schemes, the city council established or entered into partnership agreements with development interests. In many instances the city owned part of the site or one of its roles was land assembly via compulsory purchase orders (CPOs). Partnership agreements dominated the strategy to develop large sites, whilst with small stand-alone relatively simple developments, standard planning controls operated involving negotiations, application and outline permission.

Different strategies operated depending on the scale of the development and the council's perception of the importance of the site.

8.3 The development context and the Birmingham Alliance

The final elements of the development scene that need to be explored are the development interests at work in Birmingham. As in other provincial cities, property development in Birmingham is the result of a combination of national and local or regionally-based development companies. Local and regional property companies have been involved with refurbishment projects, converting 1960s office space into residential lofts, and office and retail developments that fail to meet the development and investment criteria of national property interests. One local company stands out, Richardson Developments, which has established a track record of retail, office and leisure developments. However, even Richardson's large schemes (St£50m leisure complex completed in 2001) are small in comparison to the three St£300 million-plus schemes planned or being developed in the city by national property companies.

The national companies are involved in major redevelopment projects, predominantly in partnership with the city council and, unusually, in partnership with one another through the establishment in February 1999 of the Birmingham Alliance (Hammerson, 1999). Three large property companies (Hammerson PLC, Henderson Investors and Land Securities) established this partnership agreement to undertake Europe's largest city-centre regeneration project. These companies owned substantial sites in the city and were planning to construct retail and office developments. The problem was that each development scheme might undermine the profitability of any one of the other schemes and the city would be left with vacant retail space and a problem in attracting new investment capital to the city. The locational decision of GAP, an American clothing store, provides a good example of this problem. GAP could not decide which of the new schemes would be best suited to its needs. Once the Alliance was established, GAP decided to take two 2,140 sq. m. (23,000 sq.ft.) units in two of the developments. The Alliance had removed the uncertainty behind the competing schemes and GAP was now convinced that they would be managed in a unified manner. The properties included in the partnership are the Bull Ring, the Rotunda, Martineau Square, Priory Square, Dale House, Londonderry House and the McLaren Building (Figure 8.1). The redevelopment of these sites will produce a retail-led, mixed-use scheme that will incorporate over 250,000 sq. m. (2.7 million sq. ft.) of new retailing space together with catering, leisure and residential space.

The Alliance is an umbrella agreement formed by three limited partnerships, each dedicated to one of the three development schemes: Bull Ring retail development of 110,000 sq. m. (1.18 million sq. ft.) opening in 2003; Martineau Galleries Phase I of 16,700 sq. m. (180,000 sq. ft.) opened in 2000 and Martineau Galleries Phase II planned to start in 2004. Each partner provides one-third of the finance estimated at just under St£1 billion. The sole purpose of the Alliance is to ensure that the restructuring and extension

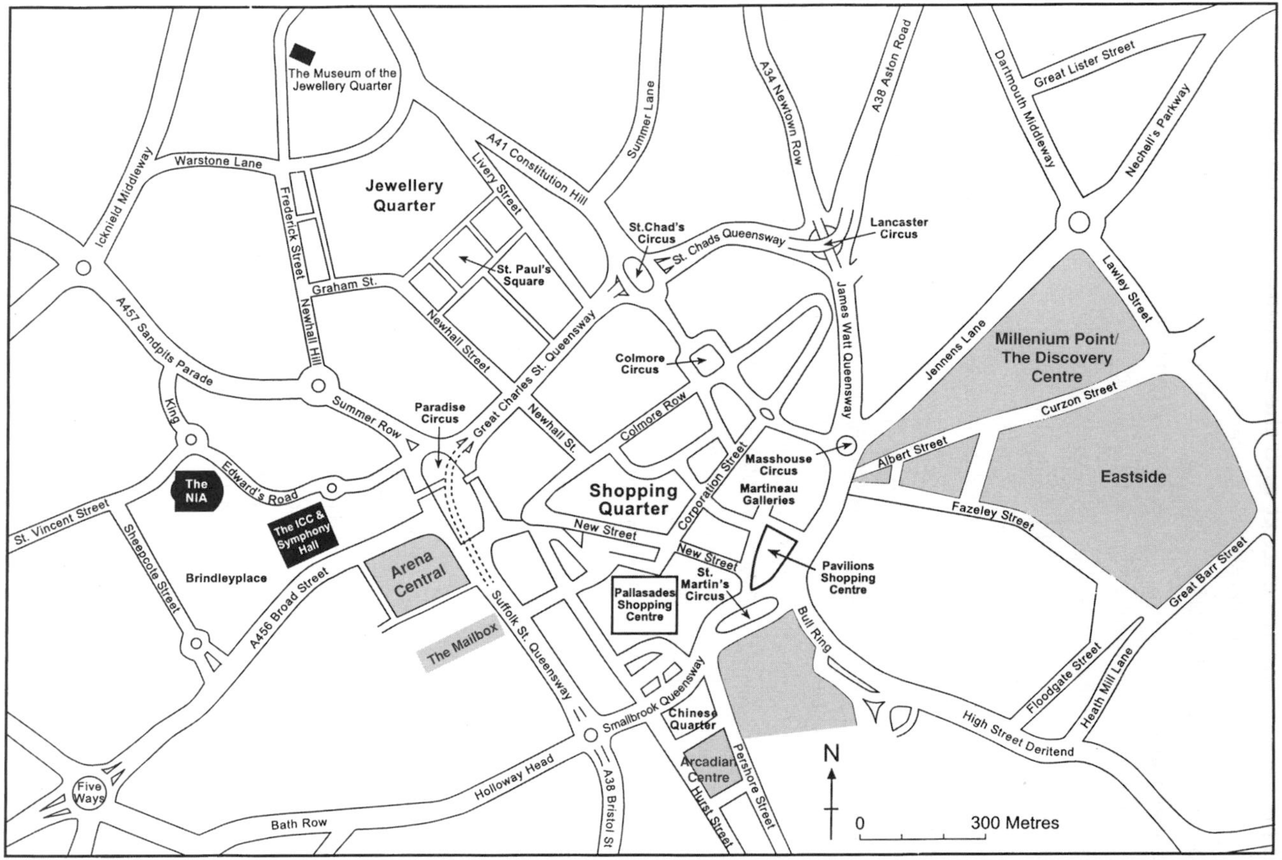

Figure 8.1 Major regeneration projects in Central Birmingham, 1987–2002

of Birmingham's retail core which is being undertaken by competing companies is phased and co-ordinated to ensure that each development and investment interest maximizes development profit and long-term investment gain. The city council is not part of the company arrangement, but has equal partner status through its extensive involvement as landowner, planning and highway authority (Birmingham City Council, 2001).The Alliance is an unusual partnership between competing private development interests that according to the chief executive of Hammerson (1999):

> ... makes good sense for all concerned. While allowing us to provide the retail facilities the City deserves, it also ensures that the developments are phased. With our partners ... we will be able to maximize the value of our interests in Birmingham.

8.4 The city of Birmingham – a city-region in transition

Since the 1960s, like many other cities in developed market economies, Birmingham experienced a substantial economic crisis. From the early eighteenth century Birmingham developed as the UK's industrial heartland and became known as the 'city of a 1000 trades' and as 'the industrial region par excellence' (Wood, 1976). The key year in the city's recent economic history was 1966; this was the high-point of manufacturing employment in the UK and of the postwar economic boom.

Between 1971 and 1993, just over half a million manufacturing jobs were lost in the West Midlands (Bryson et.al., 1996), representing 50 per cent of the total. Manufacturing employment as a share of total employment in Birmingham fell from 64 per cent in 1951, to 11 per cent in 1978, 23 per cent in 1997 and 19 per cent by 2000 (Webster, 2000) This was a period in which Birmingham experienced the crisis of Fordism – global competition, increased productivity, a developing international division of labour – and its associated deindustrialization. It resulted in massive job losses, high unemployment and a population that did not have the requisite skills for employment in the developing service sector.

During the 1980s and into the 1990s, Birmingham was left with a shrinking manufacturing base and a challenge to replace manufacturing employment with office-based jobs and to develop the infrastructure to transform the city into a major centre for service activities and business tourism. The city developed policies to restructure the economy by a process of diversification that incorporated plans for the development of professional and financial services, high-tech industries, the retention of automotive manufacturing employment and the development of a city of culture which would provide an environment that would attract service work and professional employment to the city. Like many other cities facing an economic crisis, Birmingham was attempting to manipulate its physical, social and cultural environment to transform itself into an 'attractive city' (Berg et.al., 1999); a city that contained buildings and cultural institutions that would differentiate it from competing cities and encourage the development of a city designed for the information age.

This task has not been easy. The development of an inner ring road in the 1960s left Birmingham with a small city centre; so small that Birmingham has one of the smallest retail cores of any British city. The inner ring road was also the apogee of the city council's policy to develop a transportation system designed for cars rather than people. One indication of this design flaw is the number of pedestrian underpasses incorporated into the scheme. These rapidly became spaces of fear – sites for muggings, drug abuse and graffiti – and are currently being removed. Furthermore, service firms have suffered from land and building shortages as they have been reluctant to consider offices located beyond the ring road. Second, the construction of the Ring Road involved the local authority purchasing a substantial part of the city centre. By 1971, the authority owned 93 ha. (230 acres) of the total 281 ha. (695 acres) within the central area (Borg, 1973). The inner ring road was opened in April 1971 with the city council releasing development sites for office and retail schemes on long leaseholds (Marriott, 1967).

8.5 Highbury and the city-centre strategy

The crisis experienced by Birmingham's manufacturing economy was reflected in the city centre. During the 1980s, in comparison to other British provincial cities, Birmingham had limited success in attracting investment capital. In the early 1980s, office completions reached a record low of less than 10,000 sq. m. (100,000 sq. ft.) and retail floorspace actually diminished with the closure of several department stores (Birmingham City Council, 2001). The city council realized that a co-ordinated strategy would be required. To further this objective, a draft City Centre Strategy was formulated in 1987. This strategy provided a vision for planning the city centre constructed around the identification of seven distinct 'quarters': city centre core, greater convention centre area (westside), jewellery quarter, Chinese quarter and markets area, gunsmiths' quarter, Aston triangle and Digbeth (eastside) (Figure 8.1). Each quarter had its own distinctive characteristics and planning requirements and each would be the subject of a master plan.

The final strategy represented a major shift in policy that was informed and formulated by the Highbury Initiative, two symposiums or brainstorming events convened by the city council in 1988 to develop an action programme for the city centre. This initiative was the first City Centre Symposium held in the UK and reflected the council's belief in the value of 'importing' external advice which applied the lessons, good practice and experience from elsewhere (Birmingham City Council, 2001). Representatives from Birmingham's various interest groups were brought together with planning professionals and architects from Los Angeles, Tokyo, the UK (Will Alsop, Judy Hillman, David Lock, Peter Rice and Terry Farrell) and other European countries to discuss the future of the city centre.

The Highbury Initiative underpins what has subsequently occurred in Birmingham over the last 15 years; it was the catalyst behind the reinvention of the city. To Mike Taylor, a member of the city council planning team, Highbury 'generated a broad base action programme of positive City Centre changes and a new City Centre Strategy designed to unlock the latent potential of the City Centre and tackle the physical problems left by the comprehensive redevelopment of the 1960s' (Taylor, 2000, 3–4). Eight key principles were established at Highbury (Birmingham City Council, 1988):

1. To break the concrete collar of the Inner Ring Road, reducing its influence on the city, and to give greater priority to pedestrians.
2. The development of the 'quarter philosophy'.
3. The development of a strong east–west pedestrian orientation to the city centre.
4. Birmingham's 51 km. (32 miles) of canals were to be used as the focus for regeneration schemes.
5. To focus on architectural quality and urban design.
6. The development of a 'streets and squares' approach to regeneration.
7. The introduction of American city management practices, especially the creation of associations to manage each of the 'quarters' and a city centre management team.
8. The replacement of traditional land-use zoning with a strategy that encouraged mixed-use regeneration schemes.

Following Highbury, a shared vision for the future of Birmingham was published and the Highbury vision was incorporated into the City Centre Strategy and Birmingham's Unitary Development Plan. The city council also commissioned planning consultants to produce detailed design strategies for each of the 'quarters'. These 'quarter strategies' were informed by Highbury, but operated at a different spatial scale. The scale at which planning and development occurs is extremely important. Highbury constructed a vision for the city, whilst the detailed design strategies attempted to articulate the Highbury vision to the problems faced by particular localities and quarters. The relational scalar logic (Brenner, 2001) of 'city centre' and 'quarter', 'quarter' and 'site' and 'city centre' and 'suburbs' dominates the planning and development process. Highbury subdivided the city and subsequently the quarter plans did not have to take into consideration planning issues at the scale of the city; the quarters rather than the city framed the planning process. The masterplans for each of the quarters provided an overarching framework isolated from the rest of the city. The quarter plans enabled the city to construct distinctive quarters, whilst at the same time isolating each quarter from the wider city context enabling development interests and the city council to ignore or manipulate the development process to suit the interests of private capital and the city council. Development logic comes first, and different planning scales permit the power of capital (the city council and private enterprise) to manipulate what is built and where it is built.

Highbury revitalized the city's approach to regeneration by ensuring that its future was informed by the experience of cities such as Boston and Baltimore. Highbury revealed that the city council had to operate at two levels. First, as an important landowner it should act as a major development actor by assembling sites and constructing flagship developments. Second, it should influence the actions of others (Birmingham City Council, 1988). The realization that the city council could not regenerate the city centre by working in isolation led to the creation of a number of Highbury-inspired associations. Birmingham Forward, a partnership of major professional and financial firms, was established in 1987 to promote the development of Birmingham as a commercial and financial centre, while the Birmingham Marketing Partnership, was established as a public–private partnership to promote the city as a site for business tourism.

After Highbury, part of the city's strategy was to reinvent itself as an international city which contained cultural institutions that would attract tourists and especially business tourism. Attracting people to the city was the first stage in encouraging companies to consider establishing business outlets in the city. The city was trying to act strategically to alter Birmingham's external image. The strategy operated at a number of different levels, from that of encouraging property investors to refurbish 1960s buildings to persuading Sadler's Wells Royal Ballet to relocate from London to the city by providing a theatre with facilities equal to that of a resident company (Woodcock, 1991).

The key to understanding the transformation of Birmingham's city centre is the development of top-down policies and interventions that entailed the development of a continuing partnership between the private sector and the local authority. The impetus came from the city council which invested public funds (from local and national taxation and from European Regional Funds) to provide an environment that would encourage private investment. What occurred at Birmingham is the construction of a partnership between the local state, cultural institutions and the private sector. This partnership is constantly reworked depending on the operational scale – from the macro city-level down to negotiations over individual plots.

The Highbury philosophy of stakeholder engagement has been incorporated into Birmingham's approach to regeneration. An annual meeting of key developers, landowners and representatives from business and community groups is held to discuss progress and to ensure that a regular exchange of ideas continues to occur. The result has been the redevelopment, reinvention and expansion of the city centre. This is an active process of partnership between Birmingham City Council and private development companies that has led to over St£1.5 billion being invested in development projects. The city centre strategy involved a large extension to the city centre and the replacement of poorly designed 1960s retail and office schemes. The size and complexity of the task meant that the Highbury strategy had to be phased with the focus initially being on the westside of the city and it was only in the late 1990s that attention shifted to cope with the problems faced by the city's eastside. These two regeneration areas will be explored in turn.

8.6 Westside story – Brindleyplace: 1983–2002

Birmingham's westside development (see Figure 8.1) provided a blueprint for regeneration that has informed the city's strategy for the creation of a city suitable for the information age. The strategy was inspired from a visit which the city planning committee and officers made to the USA in 1984 to explore American convention centres as a mechanism to leverage substantial private sector investment in declining areas on the back of public-sector investment. This delegation was particularly impressed by Boston's Faneuil Hall Marketplace and James Rouse's redevelopment of Baltimore Harbour. It is these schemes that formed the model for what was to occur at Birmingham, while Birmingham's own success in creating and implementing this strategy is encouraging other cities to use Birmingham as a model for regeneration. Currently, Buenos Aires is planning a regeneration scheme informed by Birmingham's experience (Johnson, 2002). Significantly, the Royal Town Planning Institute (RTPI) presented the first President's Special Award (2000) to the planners of Birmingham city council for the radical transformation of central Birmingham. The award recognizes the city's 'special contribution to planning achievement over the past 20 years … that has transformed Birmingham from a regional hub into a truly international city' (Royal Town Planning Institute, 2001). The RTPI identified six factors that contributed to Birmingham's regeneration in which the city council:

1. Led with a broad, proactive vision, its City Centre Strategy, which was based on integrated regeneration when other cities continued to focus on a minimalist statutory role.
2. Was behind a strategy to diversify the economy to include new sectors: urban tourism, international conventions, leisure and information-age businesses.
3. Developed a powerful urban design approach to create a high-quality, people-friendly environment. This has produced new streets, squares, pedestrian spaces and quality buildings.
4. Replaced traditional land-use zones with new mixed-use quarters, each with a distinctive sense of place.
5. Adapted creatively to the different public and private funding opportunities and delivery mechanisms.
6. Developed strong partnerships with local and national politicians but also with local community, business, as well as regional, national and international development and investment interests.

These six points can be illustrated by the experience of the westside regeneration scheme.

The westside story began in 1983 when the city council was exploring ways to diversify the city's economy. Part of this strategy was to designate a zone for entertainment and leisure facilities that would support a flagship convention centre located on the edge of the city centre. The site of this new zone was for two centuries a bustling mass of canal wharves, canal boats and workshops. By the 1980s it was an area of derelict land and decaying industrial buildings. The proposed International Convention Centre (ICC) (see Figure 8.1) was intended to be the largest conference venue in the UK. This was to be managed by and complement the council-owned National Exhibition

Centre (NEC) located at the airport. The convention centre was designed to stimulate and pump-prime the regeneration of this run-down area of the city. This was designed as a 'grand project' that was supposed to develop the city's role as an international convention city and to ensure that Birmingham was transformed to fit its new marketing tagline as 'the meeting place of Europe'.

The site of the ICC was designated as a Comprehensive Redevelopment Area to enable the council to assemble a 16 ha. (40 acre) site for public buildings and a 7 ha. (17 acre) site for commercial buildings using compulsory purchase orders. The plan was to create a new pedestrian-friendly quarter outside the traditional city centre core that would contain the ICC and the National Indoor Arena, an international sports centre with a seating capacity of 12,000 (opened 1994). The ICC would also incorporate a state-of-the-art concert hall designed as the home for the City of Birmingham Symphony Orchestra (CBSO), led until the 1990s by the high-profile Sir Simon Rattle. Symphony Hall was used by the city council as a marketing tool to demonstrate that Birmingham was a city of culture as well as a city of a thousand trades. The council wanted to highlight the way in which art and culture were playing a significant role in the regeneration of the city.

Between 1989 and 1992, the city council invested St£180 million in the ICC complex, St£57 million in the NIA as well as St£103 million in developing the NEC (Loftman and Nevin, 1996). Of this investment, St£163 million was raised by NEC Ltd, a public–private partnership controlled and underwritten by the council with the rest of the funding coming from the European Regional Development Fund and capital receipts from land sales. The 23 ha. (57 acre) site was cleared of its factories and buildings, leaving the Crescent Theatre (1964) intact and a boarded-up semi-derelict school listed for protection (Grade 2). The city council originally intended to develop the NIA and surrounding area itself, but changed its mind and decided to hand over the scheme to the private sector. Developers were invited to tender for the development of the 10.1 ha. (25 acre site) with the stipulation that the indoor arena should be completed by 1990 and the rest of the development by 1991. The development brief also specified a range of alternative uses for the rest of the site – mostly leisure or retail uses with offices and housing at the bottom of the list.

The subsequent development history of the site is complex, and has been reviewed in detail elsewhere (see Latham and Swenarton, 1999). All that is required here is a brief overview. Twenty-one expressions of interest were made by developers, and a short list of eight identified. Included in this list was Merlin working with James Rouse, the developer of Baltimore's Harbour, and Shearwater and Laing, the retail arm of Rosehaugh, the prominent 1980s British development company. Eventually, Rouse and Shearwater combined forces and their scheme won the competition. Merlin, Shearwater and Laing paid St£23.3 million for the 10.1 ha. (25 acre) site on a 150-year leasehold from the city with time-related development obligations linked to the construction of the NIA. They began to build the indoor arena on the least accessible and attractive part of the site. On completion of the NIA, Shearwater and Merlin created a joint company, Brindleyplace PLC, to plan and develop the rest of the site. The developers employed the architects of the ICC to develop a master plan for the area that would include a large shopping mall/festival marketplace abutting the canal, multiplex cinema, aquarium, hotel and offices. At this stage, Merlin withdrew from the partnership due to financial difficulties, leaving Shearwater in control. The decline in the property market meant that between 1989 and 1991 nothing happened on the site apart from meetings with planning officials. Shearwater, building on the experience of Rosehaugh Stanhope at Broadgate (London), decided to alter its plans for the site. At Broadgate, Rosehaugh had been successful in developing a substantial office complex away from the traditional office core of the city. Broadgate's success rested on an urban design concept that included public spaces and squares and a site with good accessibility to public transport. The developer decided to replicate Broadgate at Birmingham as the Brindleyplace site was adjacent to the city centre. The Terry Farrell Partnership was employed by Shearwater but briefed jointly by the city and developer to design a master plan that would increase the development's office space while retaining the concept of a mixed-use scheme. The draft master plan was completed in February 1991 in the same month as the city recruited a new Director of Planning and Architecture, Les Sparks. Sparks was willing to accept that the scheme should be dominated by offices, but insisted that housing was also included in the scheme to support a new national policy to bring people back into British cities.

Rosehaugh's involvement in the site should be seen as the first phase in the development of Brindleyplace. In November 1992, Rosehaugh's bankers withdrew credit facilities and the company went into receivership. Brindleyplace PLC was sold by the receivers to Argent in May 1993 for St£3m. This is an extremely important part of the westside story. Initially the land at Brindleyplace had cost about St£2.47 million per ha. (St£1m per acre), but was acquired by Argent for St£296,000 per ha. (St£120,000 an acre). This substantially altered the development equation, permitting Argent to include large public spaces and community facilities in the scheme. This increased the physical attractiveness of the development and was an important contributory factor to its eventual success.

Argent's first task was to revise Terry Farrell's master plan to bring it into line with its commercial strategy. The problem with this plan was that Argent considered it to be more suited to an 'all-in-one-go development than to a step-by-step approach' (Madelin, 1999a, 33). The new approach would enable Argent to meet the requirements of a Section 106 agreement to build speculatively the first 5,570 sq. m. (60,000 sq. ft.) of retail space and to refurbish the derelict school. A Section 106 agreement allows the planning department to grant planning permission subject to the developer's signing a legal agreement which obliges the developer to provide funds or to carry out identified works prior to construction. The rest of the site could be developed in line with market demands. The key changes included (Chatwin, 1999, 27):

1. Adjusting the plan to permit greater flexibility in the design of individual buildings. Rosehaugh's scheme had been based on large floorplates similar to those at Broadgate. The revised masterplan provided the framework for a set of buildings that were to be designed by different architects. Each building would be distinctive, constructed or faced in brick and related to each other and to the overall area in such a way as to develop a sense of place. An important constraint on the development was that rental levels were expected to be around St£215 per sq. m. (St£20 per sq. ft.). This placed a considerable design constraint on the architects and was one of the reasons behind the decision to build in brick rather than using a panel system (Partridge, 1999); brick was also more fitting for a city renowned for its red-brick architecture.

2. Inclusion of housing in a designated part of the site rather than piecemeal inclusion over the site. This would enable the scheme's housing element to be constructed first. This part of the site was sold to a housing company to finance the construction of public spaces and infrastructure.

3. Increase the size and prominence of the public spaces.

Brindleyplace was the largest mixed-use city centre development scheme in the UK. Its success lies in the ability of the city council to develop an overall design strategy for the city centre and to encourage its implementation by flagship cultural projects linked with substantial land assembly. The size of the site is an important factor in explaining the success of the scheme. Linked to this is Rosehaugh's collapse, the low cost of the land and Argent's development strategy. This strategy involved developing Brindleyplace in phases and as a mixed-use scheme containing a mix of offices in 10 buildings (102,000 sq. m. (1.1 million sq. ft.)), two new squares, shops, restaurants and pubs (93,000 sq. m. (100,000 sq. ft.)), a 240-bed hotel, the Sea Life centre (England's largest aquarium), the Ikon art gallery, theatre, 143 canal-side apartments and town houses, 34 serviced apartments, a 3,250 sq. m. (35,000 sq. ft.) health and leisure club and 2,600 car parking spaces. The total value of the development was over St£250 million.

Due to the difficulty of obtaining finance for speculative development, Argent's strategy was to pre-let buildings and develop Brindleyplace to meet market demand as it arose. The site developed in a controlled but piecemeal fashion as the developers responded to opportunities to sell land and pre-let buildings. The secret was a flexible design framework which attempted to produce a new work, leisure and living quarter of the city by providing profitable quality-designed and constructed buildings. This meant that each building was designed to suit the requirements of a particular tenant as well as the constraints imposed on the developers by property investors; financial institutions preferring buildings leased to a single tenant. Brindleyplace overcame this constraint by assembling a range of different uses, but housed in different buildings. Architects for the individual buildings had to work within the constraints of a common design framework to ensure that the scheme worked as an integrated urban design.

8.7 Eastside story – Millennium point: 1997–

Until 1997, most of the redevelopment of Birmingham's city centre was concentrated in the city's westside. It was sensible to concentrate public- and private-sector resources into one area as this enabled synergies to develop. The next major task facing the city is 170 ha.(420 acres) of semi-derelict land stretching eastwards from the city's central shopping area, including Aston University, the Science Park and Digbeth (see Figure 8.1).

In December 1999, the city council unveiled an ambitious 10-year plan to transform the eastside from a concrete desert into a quarter for learning, technology and heritage. An Eastside Partnership Steering Group, chaired by Alan Chatham, who was formerly in charge of Brindleyplace, was established to co-ordinate the creation of the eastside master plan. The partnership is between Birmingham city council and major landowners in the area: Aston University, the University of Central England, Millennium Point, Aston Science Park, Advantage West Midlands and English Partnerships.

Chatham has become an important character in the Birmingham property scene. He was in charge of Brindleyplace and, on its completion, began to look for new development opportunities in the city. He realized that Birmingham's largest building, the postal sorting office, was going to be sold, and tried to persuade Argent that converting the building as a mixed-use scheme would be a profitable undertaking. Argent was unconvinced and Chatham left to establish his own development company. He purchased the Mailbox for St£3 million in 1998 (and St£1 million for adjacent waterfront buildings) and rapidly sold the building's air-rights to a house-building company, Crosby homes. This sale provided him with some cheap development capital. Chatham converted the building to contain two hotels (300 rooms in total), 15,850 sq. m. (170,000 sq. ft.) of quality office space, 9,290 sq. m. (100,000 sq. ft.) of retail space and a similar area devoted to restaurants and a health club. This was a highly successful conversion as Harvey Nichols, the upmarket department store, acquired two of the units, and the British Broadcasting Corporation is relocating its Birmingham Pebble Mill studios to the development. Chatham's St£4 million purchase was converted into a mixed-use scheme valued in excess of St£125 million.

Whilst being formally announced in 1999, the redevelopment plans for the eastside began in 1997 and reveal that the eastside story is going to be similar to that of the west. In 1997, with the rapid onset of the millennium, Birmingham city council decided to establish a flagship development in Digbeth, similar to the ICC and Symphony Hall. It decided to create an educational hub in the eastside by investing St£111.6 million in creating a Millennium Point Centre. The funding was a mix of European (St£23.8million), central and local government financing, together with private-sector sources and St£50 million donated by the Millennium Commission, a fund established from national lottery funding for projects to celebrate the millennium. The concept was to relocate the city's Science Museum to a specially-designed building that would include a Technology Innovation Centre run by the University of Central England to provide engineering and computer technology courses to business, a University of the First Age that would provide a multi-sensory learning environment for young people and the Hub, a cluster of places to eat, drink, shop and relax. Millennium Point was opened in 2001 and is being used as the development node to attract private-sector investment to the area.

While the Highbury Initiative had been associated with a strategy involving the relocation of Sadler's Wells Royal Ballet to Birmingham, eastside also boasts the relocation of a similar high-profile cultural institution from London. The Royal College of Organists (RCO), the world's leading institution and professional body for the promotion of organ and choral performances, has been persuaded to relocate to Birmingham's eastside. The college is acquiring the premises of the former Curzon Street Station, a Grade I listed building, to create an organ and choral education centre containing a library, recital hall, practice rooms and conference facilities. The move reflects Birmingham's developing artistic community that is beginning to operate as a cultural agglomeration. Close links are being established between the RCO, Symphony Hall and the Birmingham Conservatoire, the city's music college. In Birmingham, collaboration between the various cultural institutions is possible and even welcome, for example between Birmingham Royal Ballet, the CBSO and the city repertory theatre.

The Eastside Partnership commissioned the architects HOK International to develop a masterplan for 40 ha. (100 acres) of eastside. This was published in February 2002 and provides a framework for a major mixed-use development that will include the continued creation of a new cultural, learning and technology quarter. Like Brindleyplace, the plan is intended to attract commercial investment by providing a framework of new complementary neighbourhoods that are linked to an extensive parkland network. A new 3.6-ha. (9-acre) city park will be created along with new squares and the refurbishment of old buildings and the introduction of landmark developments. Part of the plan involves removing Masshouse Circus, an elevated part of the Inner Ring Road and associated roundabout, to break the 'concrete collar' that the road imposes on the development of the eastside. Masshouse Circus is being replaced by a network of tree-lined streets with the city council releasing the land it owns in the vicinity for private-sector development. The eastside project is the latest stage in the implementation of the Highbury Initiative and reflects a continuing emphasis on urban design, city centre involvement in encouraging development and partnerships between the public and private sectors.

8.8 Birmingham's dilemmas

Birmingham's redevelopment has been associated with four unplanned consequences or dilemmas that need to be taken into consideration during the development of the eastside.

First, the regeneration of Birmingham is tied into an economic development strategy to diversify the city's economy away from metal bashing towards service employment. The problem is that whilst the city centre has been expanded and its infrastructure and built stock transformed, the impact on the city's employment has been problematic for two reasons. First, many of the new buildings have encouraged displacement or intra-urban relocation that has produced voids in 1960s 'obsolete' office buildings. The success of Brindleyplace has therefore produced problems for other parts of the city. Second, the strategy to develop Birmingham as a convention centre improved the image of the city and led to Birmingham's hosting the European Union summit meeting in 1992, the G8 gathering and Eurovision Song Contest in 1998. The city council claims that Birmingham is now the fourth most visited city in the UK (after London, Edinburgh and Glasgow) with over 22 million visitors a year, but 70 per cent of this is business tourism. Tourism supports 27,350 jobs in hotels and restaurants (Birmingham City Council, 2000). However, most of the jobs provided in the new developments that benefit disadvantaged communities are low-paid and insecure. In 1991, 42 per cent of the 275 permanent jobs at the ICC were in cleaning, catering and security (Loftman and Nevin, 1996) whilst the city's six new hotels created 700 new jobs, but the majority of these posts are in poorly paid occupations.

Second, Brindleyplace, and the city council's investment in cultural infrastructure, has benefited the city's middle classes by providing them with work, living and consumption spaces that implicitly exclude the lower paid and, in many cases, minority ethnic groups. The Bull Ring redevelopment is an excellent example of this form of city restructuring designed for urban élites. The old Bull Ring centre had become a location for cheaper stores intended to meet the needs of the city's poorer groups. The new Bull Ring is designed for high-profile expensive stores; it is going to provide a designed environment to maximize conspicuous consumption. This is not to criticize the workings of property capital that must seek to maximize profits, but is rather a critique of city council policy that has favoured these types of high-profile regeneration schemes. By 1999, the city council calculated that 33 per cent of the city's employees are 'knowledge workers' (Birmingham Economic Information Centre, 2000). Other research has shown that Birmingham has the largest professional services sector in the UK after London (Daniels and Bryson, 2002). The question is really 'for which group is the city being redesigned or replanned?' The answer to the question is complex but, ultimately, Birmingham is being replanned in the interests of property capital and for the interests of the city council in its attempt to position itself in the competition amongst cities to attract forms of foreign and domestic direct investment. The result has been the creation of a built environment that is designed to meet the needs of an urban élite or to provide spaces designed for white cultures of consumption and to exclude or ignore the disadvantaged as well as minority ethnic groups from these spaces.

Third, the city council's dual role as catalyst and developer–investor comes at a cost; it displaces investment and management time that could be spent addressing other problems. The focus has been on the most visible prestigious schemes. Birmingham's planning department has a poor record of dealing with small-scale planning applications and it is now fast tracking major planning applications at the expense of delaying other planning decisions relating to small infill schemes or factory extensions. This is one of the unaccounted costs of the strategy. There is a more visible cost; the diversion of investment from the suburbs to the centre. Loftman and Nevin (1996) demonstrate that the scale of investment in flagship projects distorted the city council's expenditure programme from 1986. They show that, between 1986 and 1992, public-sector investment in Birmingham's housing stock was 28 per cent below the average for English local authorities and that capital expenditure on education in the city fell by 60 per cent compared with increases in other cities such as Manchester and Liverpool. The point is that the flagship projects funded by public funds to encourage private sector investment negatively impacted on disadvantaged groups and areas in the city. This is a substantial problem as over two out of every three children in the city are being educated in sub-standard buildings and over St£100 million needs to be spent on school buildings in the city (Loftman and Nevin, 1996).

Fourth, Birmingham's redevelopment has stressed the importance of mixed-use schemes. The squares and roads of Brindleyplace were funded by the sale of a 1.2 ha. (3-acre) site for houses. Initially, this provided Argent with a major advantage – cheap, if not free, capital to underwrite those elements of the scheme that did not directly generate profit. However, the homes were completed and occupied before Brindleyplace was completed and the developers experienced a 'my home is my castle' or 'not in my backyard' effect. The new high-income residents delayed one phase of the project by objecting to the planning application. This cost Argent at least St£1 million and led to the loss of potential tenants because buildings were not finished to the planned timetable (Madelin, 1999b). The residential components of a mixed-use scheme can over time remove or distort many of the features designed to produce a lively urban environment. The housing component of Brindleyplace (King Edward's Wharf Lofts) was marketed by highlighting the advantages of living close to live music venues including the Fiddle and Bone, a live music bar established by two members of the CBSO. However, when the live music licence came up for renewal, residents objected on the grounds that the music exceeded sound levels for residential areas and the pub lost its licence.

8.9 Conclusion

The Highbury initiative is an excellent example of strategically-informed city centre planning. The success of Birmingham's regeneration is founded on an ongoing dialogue between private-sector urban development interests and the city council. Other stakeholders, for example local residents and ecological or environmental groups are marginalized in this dialogue. The city council, in consultation with key stakeholders, established a planning framework for the city. Sub-plans, created by the local authority or private–public partnership, are created for individual quarters, while masterplans are created by developers who have acquired substantial sites. The smallest scale of planning intervention involves negotiations over individual plots, but negotiations are restricted to design features of individual buildings as the various agreed planning frameworks determine the type and nature of the development required for each site. This nesting of scales of planning intervention is one of the lessons that come from Birmingham's attempt to recreate itself as a city for the twenty-first century. The ability to transfer responsibility for the development of commercially viable master plans placed within a wider city council controlled framework ensures that buildings are constructed to meet the constraints imposed by potential tenants as well as investors. All this suggests that Birmingham's regeneration has been manipulated for the interests of private capital, but some of this manipulation has been undertaken by the city council as part of its policy of proactive planning. Furthermore, the Birmingham Alliance demonstrates that development interests are willing to work together when they realize that working apart might reduce or even undermine the profitability of their individual developments.

The Highbury Initiative is underpinned by four city council strategies, some reflecting national planning policies. First, many of the development schemes were funded or inspired by public-sector investment that increased private-sector development profitability. The flagship schemes altered the image of parts of the city and, at the same time, solved infrastructural problems related to the inner ring road. Second, the city council has played an active role in securing funding from a variety of sources (European Regional Development Funds, Single Regeneration Budget (UK) and the National Lottery) used to finance flagship developments and improvements to the city centre. Third, planning has used Section 106 agreements to acquire planning gain in the form of new community facilities, the refurbishment of derelict listed buildings and the development of public buildings. Fourth, the city has actively used compulsory purchase orders, often in partnership with private-sector developers, to assemble sites suitable for major development projects. The council has also used its powers in a number of cases to close roads that would impede large-scale redevelopment.

Overall, the lesson from Birmingham is one of a city trying proactively to use urban design to shape the built environment through policies, programmes and guidelines rather than detailed plans. The detail is constructed in partnership with private capital and the strategy is creating a mixed-use urban environment for the twenty-first century. However, this raises the question of whether too much emphasis has been placed on the requirements of private capital at the expense of creating inclusive spaces that are attractive to Birmingham's various multi-cultural communities and disadvantaged groups.

9

Growth machines and growth pains: the contradictions of property development and landscape in Sioux Falls South Dakota

Carrie Breitbach and Don Mitchell

9.1 Introduction

Many people in the USA – and around the world – have heard of Sioux Falls, South Dakota, even if they do not know exactly where it is. Millions of people receive letters from Sioux Falls offering them a Visa card – at some low introductory rate – from Citibank (one of the largest credit card companies in the world). Millions of others pay their monthly credit card bills to addresses in Sioux Falls. For anyone over 40, however, Citibank is more closely associated with New York, where it began life in 1812 as the City Bank of New York. By 1930, the City Bank of New York, with affiliates in Asia and Europe, was the largest bank in the world, helping to make New York City the primary global financial center. Still headquartered in New York, it is now called CitiGroup (the corporate parent to a diverse array of banking and other financial services businesses) and operates in 102 countries around the world. In 1981, Citibank shocked New York and a number of other suitors by relocating its credit card services to Sioux Falls, a provincial city of just under 124,000 in the north central plains of the US more known for meatpacking than for New York-style financial acumen. In the words of *Forbes* magazine writer, Citibank's relocation 'put little South Dakota back on the map' (Merwin, 1983, 85). The relocation did not just happen but was induced, or cajoled, by a coalition of property and other economic interests in Sioux Falls as part of a profound reshaping of the property landscape of this small city.

The first site of Citibank's presence in Sioux Falls was a leased space downtown in the 70-year-old Western Surety building but, after a year, the headquarters were moved to a $25 million complex on 10 acres in an industrial park in the northeast of the city. The industrial complex is one of several created by the Sioux Falls Development Foundation beginning in 1970. Today, Citibank is surrounded by other businesses, including Gateway Computers, Hutchinson Technology Inc., and the Wells Fargo credit card center, in a green and neatly maintained expanse of buildings that differ markedly from the old pink quartzite buildings of downtown, or, for that matter, from the site of the previous number

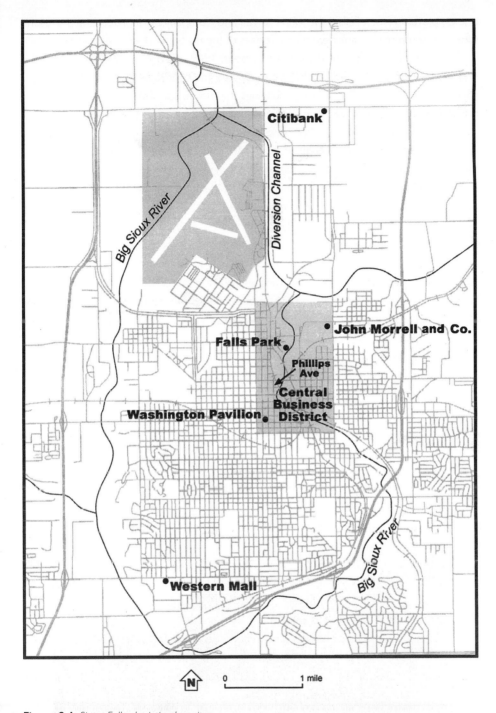

Figure 9.1 *Sioux Falls, depicting key sites*

one employer in Sioux Falls – the John Morrell and Company pork processing plant. John Morrell is south of Citibank near the waterfalls of the Big Sioux river that gave Sioux Falls its name, where the edge of town used to be in 1909 when Morrell's opened (Figure 9.1). The complicated buildings that make up the meat processor are surrounded by parking lots, railroad tracks and small buildings that house social services. The differences between historic downtown Sioux Falls (Figure 9.2), the site of the meatpacking plant and Citibank's California-style office parks are visual testimony to the strides a particular set of people and institutions have made to change Sioux Falls from a mid-sized Midwestern agricultural town to a nationally recognized urban growth success story. In the words of a *New York Times* writer, Sioux Falls' leaders managed to change the city from one of 'prime cuts to prime rates' (Brooks, 1989, 17).

The Sioux Falls Development Foundation has been at the forefront of this transformation. Begun by the Sioux Falls Chamber of Commerce in the 1950s, the foundation has emerged as the leader of the city's redevelopment. Since 1970, the Development Foundation has been buying and building infrastructure on what was once farmland to the northeast of the city, to prepare it for new and relocated businesses. As of 2000, the foundation had six industrial parks in the northeast and had recently purchased land for another in northwest Sioux Falls. As it has been purchased and developed, the land has also been continually annexed by the city so that city services can be extended to the parks and so that its tax revenue can be captured by the city, rather than by competing jurisdictions.

Figure 9.2 *A historic pink quartzite building in downtown Sioux Falls*

Through the industrial parks, the Development Foundation has shifted both the type and the spatial structure of the city's economy, shifting major economic players to the periphery and leaving the core for new uses. A network of other groups has therefore emerged to revamp downtown and other parts of the city. Among the most prominent have been Forward Sioux Falls, primarily a fundraising and investor-relations program, and Downtown Sioux Falls, an active and expanding program that has served to reshape the center of the city to accompany its new economic base. Much of the work of Downtown Sioux Falls has been to compensate for the damage done to the city centre by federally-funded urban renewal projects since the mid-1960s which replaced popular pink quartzite buildings with modernist office buildings (Figure 9.3) and provoked resident protests, suggesting the emotional investment some have in the look of the city. In a pattern of development familiar to many cities across the US that became subject to urban renewal programs, the revamping of downtown was accompanied by heightened development along the city's suburban shopping corridor, the east–west expanse of 41st Street. Western Mall (see Figure 9.1), the first major indoor shopping mall in Sioux Falls, was built in 1968. After city planners made use of federal urban renewal funds in 1971 to clear out older buildings downtown and build a banking, insurance and public service centre, they tried to lure shoppers back from the new suburban malls by creating a downtown pedestrian mall in 1973 and 1974. By all accounts, this was a failure and the street was reopened for vehicle traffic in the 1980s. Downtown Sioux Falls was therefore established in connection with the Development Foundation in order to revamp and lure new investment into the downtown area. The program and its major offshoot, Main Street Sioux Falls, begun in 1988, have been funded through the creation of a Tax

Figure 9.3 Downtown office buildings

Increment Financing District.[1] Downtown Sioux Falls lists among its accomplishments the creation of the Downtown Economic Development Incentive Fund, the organization and sponsorship of a number of street festivals, a marketing program and efforts that have led to an increase in downtown occupancy rates from 60 to 90 percent since 1990. The businesses and facilities that have come to occupy the downtown area include the new Washington Arts and Sciences Pavilion, two theatres, two National Register historic districts (one containing a recently opened local brewery and restaurant),[2] and a series of specialty shops along Phillips Avenue, the site of the failed pedestrian mall of the 1970s.

Among the most recent of the planned redevelopment efforts is the Phillips to the Falls Project which, with the help of a new Tax Increment Financing District, will connect the revamped downtown with the also refurbished Falls Park (Figure 9.4). The plan involves the acquisition of 'privately held, underutilized, unsightly Brownfields parcels in the river corridor of north downtown' (Seten, 1999). This project will complete the reshaping of Sioux Falls' primary consumer and entertainment areas since the arrival of Citibank by removing Pitts Inc. scrap salvage yard, which has been in operation for 78 years and 'effectively discourages economic investments in neighboring properties' (Seten, 1999).

The reorganization of Sioux Falls' northern boundaries for businesses, and the renovation of its downtown for those who come to work at the new businesses have been accompanied by a boom in new housing construction, especially in the city's most expensive southern suburbs (Figure 9.5). Traffic is becoming more of an issue, as

Figure 9.4. *Falls Park*

Figure 9.5 *A home in Sioux Falls' expensive southern neighbourhoods*

suburban shopping districts and a new multiscreen cinema complex have opened around the city's interstate loops.

The effect of the multiple groups and programs in Sioux Falls over the past 25 years has been to reshape most of the city into a place that is most unlike any other South Dakota town (and despite the presence of Citibank, not at all like New York City either). Sioux Falls residents now have suburban housing estates, complete with their own suburban grocery stores and strip malls, office park jobs that have made the interstate loops around the city a necessary part of a daily commute, and a downtown entertainment and consumer district newly revamped and nearly wiped clean of any visible contradictions to the economic prosperity that has contributed to the success story of Sioux Falls. On the weekends, Sioux Falls' burgeoning middle class can mingle with tourists at the redone Falls Park, free from the harassment of the 'transients, drug dealers and delinquents' that, according to a City Planning Office worker, used to frequent the park (Seten, 1999).

Property development, and the economic development that is both the impetus for and the result of property development, has brought with it a radical transformation of Sioux Falls' landscape over the past 30 years. Not only has the economy of the city been transformed, but so too have its appearance and meaning, both to residents and 'outsiders'. While the redevelopment of Sioux Falls' property landscape has been nothing less than spectacular, it has not been without problems or contradictions. Landscapes are always sites of struggles and contradictions (cf. Mitchell, 2000, Chapters 4 and 5), and the development process is never entirely smooth. Rather, it is better understood as the attempted imposition of a particular *vision* of what the city landscape should be – and who it should be for. In this chapter we will explore the property development process in Sioux Falls by paying special attention to the contradiction between the development of the city as a set of parcels of property, and the development of the city as a complete landscape. We will do so for two reasons. The first is to show how understanding the relationship between the city as *property* and the city as *landscape* provides an insight into whom development is for, and who is excluded (or whom it is hoped such redevelopment will exclude). The second reason, following from the first, is to suggest that as successful as Sioux Falls' economic and property development has been, it has come with clear social costs. These costs may appear only as 'growing pains' but are really deeper and include the systematic exclusion of some people – the poor, youth and the homeless – from both the benefits of development and from the very spaces of the city that have resulted from this development.

To explore both the relationship between landscape transformation and property development and the social costs it creates, in the next section we will examine in more detail why it is important to link understanding of property development to landscape transformation. We pay particular attention to the dynamics associated with growth – or more accurately capital accumulation – and the contradictions that are a necessary part of it. These contradictions have to be continually negotiated by the various institutions that have arisen – institutions that can collectively be referred to as a 'growth coalition' – to induce and manage economic growth in Sioux Falls. The third section turns to the specific means by which Sioux Falls has managed these contradictions and to the social costs associated with them. In conclusion, we suggest that as seamless as the growth coalition in Sioux Falls would like its landscape and property development to be, it will always necessarily be uneven and that this very unevenness means that the social costs of development can only ever be displaced, never entirely eliminated.

9.2 Landscape, property and growth in Sioux Falls

Paul Knox (1993, 28) has written that 'although urban space is produced and sold in discrete parcels, it is *marketed* in large packages'. That is to say, in urban development and redevelopment there is an intimate relationship – and tension – between urban space as *property* and urban space as *landscape*. Urban property is a specific piece of land, or a specific set of improvements upon land, for which there are an associated set of property rights. A property right can be defined as 'an *enforceable* claim made against others to the use or benefit of something …' (Blomley, 2000, 651; see also Ely, 1992). These rights may not be absolute, but they do define how and by whom property may be used. They also determine who receives the benefits – the fruit of increased value – of any improvements to the property. Within capitalism, and especially in the USA, a property regime has developed that begins from the assumption that owners of property have a presumptive right to do as they will with their property, provided it does not adversely affect others. Zoning and other restrictions on this right have largely been accepted as a means of protecting the property rights of neighbours, as much as a means of protecting some common good.[3]

An important difference between landed, immobile property and other kinds – such as a person's possessions – is that the former exists only as part of a larger whole, a whole landscape of properties. What occurs on neighbouring properties can have profound effects, positively or negatively, for any particular parcel or property. An owner of landed property, then, has a vested interest in the properties of her or his neighbours in a way that an owner of a stereo, say, does not have an interest in her or his neighbour's stereo. If you destroy your stereo it does no harm to me. If, by contrast, you allow your landed property to become derelict, it can have a significant detrimental affect on my neighbouring land. Similarly, if you make extensive improvements to your land or the buildings on it, I may materially benefit, even though I do nothing to improve my property. So, on the one hand, you have a presumptive right to do as you wish with your property. On the other hand, I have a deep interest in limiting and shaping your right, even as I may wish to keep you from shaping my own exercise of property rights.

In this sense, urban space as property stands in partial contradiction to urban space as landscape – as a suite of linked and connected properties, existing as a social whole (see especially, Blomley, 1998). The term 'landscape' connotes the complex interaction of two related phenomena.[4] On the one hand it indicates a geographical area, or more specifically, an association or arrangement of buildings, roads, lawns, trees and so forth. In this sense, landscape is a built form – what Harvey (1977, 1982, 1989) calls a 'built environment'. It constitutes what Carl Sauer (1925) long ago labeled a 'morphology', an areal association of things-on-the-land that has a specific shape and structure. On the other hand it indicates a certain kind of representation, or a 'way of seeing' the world (Berger, 1972; Cosgrove, 1985, 1998). In this second, representational sense, the landscape works as a complex symbol of the desires, needs and histories of the place and people of which it is a part (cf. Meinig, 1979; Duncan, 1990). It is this second sense of landscape that Knox means when he says that urban space must be marketed as a package: the landscape must be made to represent that which investors desire. In the case of contemporary property development, that vision is increasingly one of seamless and rapid economic growth as the primary – indeed the only – rationale for development. The landscape must come to stand for both the success and the potential of that growth and, by definition, must therefore exclude uses and people (such as the homeless or poor alcoholics in public space) who seem to contradict that vision.

In capitalism, both urban property and the urban landscape (as well as, of course, the people who are a part of them) must be made to serve as the foundation of continued capital accumulation (Harvey, 1978). This is no easy process, and it is riddled with contradictions. One of these contradictions is that 'fixing' capital in the built environment puts it at considerable risk: risk of devaluation, of obsolescence, of outright destruction, or as indicated above, risk that your neighbours may, while advancing their own self-interest, undermine yours. That there will be social impacts with the restructuring of the landscape is inevitable, since capital is fixed not only in the built form of the environment, but also in a set of social relations. Thus, the production of a landscape is dynamic in such a way that some owners of property inevitably lose. As Harvey (1978, 124) puts it:

> Capital represents itself in the form of a physical landscape created in its own image, created as use values to enhance the progressive accumulation of capital. The geographical landscape which results is the crowning glory of past capitalist development. But at the same time it expresses the power of dead labour [labour power already transformed into material things] over living labour and as such it imprisons and inhibits the accumulation process within a set of specific physical constraints. And these can be removed only slowly unless there is a substantial devaluation of the exchange value locked up in the creation of these physical assets.

The protests over the destruction of Sioux Falls' signature pink quartzite buildings associated with urban renewal in the 1960s, like the more recent concern over what to do about the privately held, underutilized, unsightly Brownfields and the scrap yard near downtown that together seemed to be deterring investment, are evidence that the dynamic reshaping of landscape with the fluctuations of capital is rarely without social conflict and the exercise of power. This conclusion is decidedly contradictory to the image of unhindered potential for growth that Sioux Falls' leaders wish to project to developers.

Thus the successive phases of Sioux Falls' redevelopment are also evidence of a particular social desire to make over the city as a certain *kind* of landscape, a certain morphology that will properly represent the city as a dynamic, high-tech, financial services and entertainment centre. Property development, and the associated landscape transformation that is both a precondition and result of that development, thus requires what Harvey (1982, 31) calls a 'class project', even as this 'class project' is always in competition, and often in contradiction, with other (intra-class) interests both in and beyond Sioux Falls. In basic terms, this class project is one of capturing for the property and development interests of Sioux Falls some high proportion of regional and global investment.

9.2.1 Property and the growth machine

Sioux Falls is in competition with any number of other regional centres and all those places – Chicago, Dallas, Denver, Sioux City, Columbus or Charlotte – are seeking to attract the relatively mobile capital associated with financial services. It has, like these other places, therefore assembled a 'growth coalition' and the city has become a 'growth machine'. In his classic article 'The city and growth machine', Harvey Molotch (1976, 310) argued that:

> the desire for growth provides the key operative motivation toward consensus for members of politically mobilized local elites, however split they might be on other issues, and that a common interest in growth is the overriding commonality among important people in a given locale – at least insofar as they have any important local goals at all.

In other words, 'the very essence of a locality is its operation as a growth machine' (Molotch, 1976, 310). Molotch's statement of essence, as well as his emphasis on what he deems the 'important people' and the 'important local goals' reveal a conception of place as an entity led by an élite focused on growth, that has proved quite resilient over the years. The reason this seminal work received so much attention is that it recognized a shift in the focus, even the 'essence', of a city from a centre of services and provisions for residents to an organization, or machine, controlled by a select group of power-wielding people, designed for the overriding, indeed singular, purpose of economic growth. The city in this view has become a machine for the accumulation of capital.

Molotch's seminal work 'set in motion a vital research agenda that now extends across disciplinary boundaries, and embraces the broad range of urban studies' (Jonas and Wilson, 1999, 4). Despite numerous changes since 1976, Jonas and Wilson argue that the growth machine thesis has persisted 'because the basic issues that the thesis addressed – issues of growth, local economic development, and of who promotes these – remain central to the politics of cities and, indeed, of many places in the global economy' (Jonas and Wilson, 1999, 4). These issues are in fact at the centre of the changes in Sioux Falls.

A key tenet of Molotch's growth machine was its orchestration by a land-based élite, in whose own interest was the continuation of an increased return on rent and a secured basis for the perpetuation of growth. This foundation was achieved by the élite through its influence on local governmental figures and agencies – those in charge of setting regulations and legislation that had the potential to hamper or to entice investors and to allow or forbid land uses that encouraged growth. Importantly, the relationship between local government bodies and the landed élite was a co-dependent one, since the city's political leaders benefited themselves from a profitable and growing city. This relationship of co-dependency helped to ensure the perpetuation of the growth machine.

Though the co-dependent relationships between the business élite and the local government present a strong source of consensus on how city affairs are run to encourage growth, a second element is necessary for the maintenance of a growth machine: the manufacture of a 'we feeling' that will ensure that no alternative configuration of the city will arise to topple the growth machine (Molotch, 1976, 311). Molotch argued that a sense of community solidarity is promoted through government-funded 'boosterish' activities like sports teams, promotional ads, Chamber of Commerce publications, public school programs or other city events. Such activities enlist the general population who may not feel the direct effects of the growth machine, for certainly it is not an evenly distributed growth, to join in a celebration of the growth of the city (and implicitly of the legitimacy of those who have produced and will benefit from that growth). The manufacture of a 'we feeling' involves promoting and mobilizing a particular 'way of seeing' the city, and thus the growth machine is active in the construction and dynamic manipulation of landscape in the representational sense as much as in the physical sense. The growth machine itself (local government and the landed élite) may not be directly responsible for all the civic boosterism that occurs, but it 'mobilizes what is there, legitimizes and sustains it, and channels it as a political force into particular kinds of policy decisions' (Molotch, 1976, 315). This ideology that supports growth within the community may also be used to enhance the competitiveness of the city, since, as Molotch recognizes, virtually all cities in North America (where he based his work) are similarly configured for growth, generating competition between places for the types of property uses that will increase land value.

The goal of the growth machine, as a specifically located class project that nonetheless seeks to elicit widespread popular support, is obvious enough: economic growth. In the words of one Sioux Falls Development Foundation employee, the 'sole purpose [of the Development Foundation's efforts to reshape Sioux Falls] was to increase jobs and businesses in Sioux Falls'.[5] Yet this statement belies the complex processes and outcomes of property redevelopment as comprehensive as that which has taken place in Sioux Falls. That the redevelopment of Sioux Falls has indeed been comprehensive is evident in the city's chart-topping population growth rate[6] and other demographic changes,[7] the dramatic shift in its prime employer from a meatpacker to banking and health services companies and media accolades.[8] Not coincidentally, given that the Sioux Falls growth machine, like its counterparts elsewhere, is rooted in landed property interests, the Development Foundation's desire to increase jobs and businesses required a comprehensive overhaul of the entire city. To put that another way, economic development in Sioux Falls is firmly rooted in *property* redevelopment, but it is linked to a process of what could be called *social* development, a development marked by changes in laws and rules regulating the use of public spaces, the organization and location of social services, and in the meaning and physical appearance of the urban landscape.

9.2.2 *The growth machine in the urban landscape*

These social developments are necessary to the kinds of changes in property – and hence the landscape – that Sioux Falls' leaders hoped to create, in part because the ability of any new business to thrive and generate revenues for the city depends upon the nurturing of a labour force appropriate to the city's changing place in the local, regional, national and international divisions of labour, and also because city leaders use the image of the city and its residents to recruit businesses to its newly developed properties. The urban landscape is critical both to the construction of the 'we feeling' upon which local cohesion is forged, and to the national or global image that the city projects to potential investors and in-migrants.

In discussing the decision First City Bank made to relocate from Texas, bank president David Elgena said that:

> a major reason for the long-term commitment to Sioux Falls ... was the quality of the workforce the operation could attract – including wives of farmers seeking to augment a dwindling farm income and graduates of the many small colleges here ... Residents ...

> have a strong work ethic. Local families are mostly of German, Dutch and Scandinavian
> descent. The explosion of jobs has even begun to attract young farm women who had
> been drifting to the major cities as nannies.
>
> (Brooks, 1989, 19)

It can be argued, in fact, that the availability of such a workforce is itself, in part, a *result* of property development and landscape change. Indeed, the Sioux Falls Development Foundation makes exactly this argument. On the 'Membership FAQ' (frequently-asked questions) section of its website, the Development Foundation, which is primarily a property development agency, poses the question, 'With unemployment so low, doesn't economic development make it more difficult for me to find and retain employees?' and answers it by asserting:

> The growth of our job market is one of the reasons unemployment does traditionally
> stay low in Sioux Falls. However, the driving force behind the population growth in our
> community has been the strength and upward trend of our job market, and the growth
> of our business community. In other words, opportunity brings population. It could be
> argued that without the growth in our economy during the past decade, the workforce
> challenges we see today might be greater.[9]

The Foundation makes clear that its goal is to create a cycle – hopefully unbroken – of accumulation. This cycle is one that begins with the development of property; such development attracts new capital; new capital creates new jobs, which in turn attracts an in-coming labour force; but it also creates tax and development pressures on surrounding farmland, which have the effect of making farming in the region an ever-more marginal enterprise. This in turn releases land for commercial and residential development *and* induces 'wives of farmers' to enter the clerical work force, assuring that the labour pool is continually replenished; and so forth.

The result is a new landscape that fully comes to represent successful growth – it seeks to exude prosperity. But it is nonetheless a contradictory landscape. In the first place, since the cycle of accumulation and transformation rests on property, on its transformation and its development, it clearly benefits property owners – the property interests that form the backbone of the growth machine. But it does not necessarily represent the interests of *all* property owners. Consider, in this regard, the salvage yard operator who has no interest in selling or abandoning his property and who has to be persuaded or forced to do so. The 'class project' of property transformation, therefore, has to negotiate different interests, different levels of social power and different needs among its constituents.[10] To make these negotiations and to represent particular interests in different places, specific mediating institutions therefore arise which seek to give voice and substance to different interests while still ascribing to the overall project of property and economic development. In Sioux Falls, these interests are organized by the Sioux

Falls Development Foundation, which has been mostly concerned with developing the industrial parks and housing tracts at the periphery of the city, and by Downtown Sioux Falls which has taken on the role of co-ordinating the redevelopment of a downtown largely abandoned or by-passed during both the urban renewal period of the 1960s and the suburban expansion of the 1970s–90s.[11] Where the Development Foundation has used tax abatements and the trade in real estate itself to advance its development agenda, Downtown Sioux Falls has relied primarily on tax-increment financing.

The use of tax-increment financing makes obvious the contradiction that exists in the relationship between property and landscape. This contradiction is best understood in reference to the transformation of *public space*, which is such a key part of any city's landscape and any city's redevelopment efforts.[12] Public space is a specific form of property, property that is owned by the (local, regional or national) state and held open for public use (see Mitchell and Staeheli, forthcoming). It is typically owned in the name of the public and the general presumption is that it is open to all. But since public space is *property*, it (like any property) is defined by a specific set of 'rights-to-use' (see above) and a complementary 'right to exclude' unwanted uses and users. Tax increment financing assumes that there will be a general rise of property values across a district. By overseeing the redevelopment of downtown Sioux Falls as a whole landscape, the Downtown Sioux Falls program prepares the ground for specific investments which will presumably lead to increased property values. The tax increment thus gained repays the bonds and other instruments used to finance the preparation for investment.

One of the key aims of tax-increment financing by Downtown Sioux Falls has been on revitalizing the public spaces of the downtown core, in essence giving the downtown landscape a 'make-over'. Sioux Falls, like many other North American cities in this regard, views the rehabilitation of public space as critical to development success. Such public space revitalization has included the provision of street furniture, the reconditioning of sidewalks (often remaking them in an image of some historic moment), and the redevelopment of downtown parks (cf. Davis 1990; Low 2000; Zukin 1995). That is, Downtown Sioux Falls, through its focus on public space, reconfigures the landscape so that it is attractive to investment. It exercises a right to shape or remake (public) property, to exclude some uses (and some people) and to encourage others. But it can do so only to the degree that its actions will lead to a *general* rise in property values – that is a general rise in *private* property values – throughout its district. To do otherwise would not only threaten its own reason for existence, but also its very *means* of existence: increased tax revenue.

Or to put this in slightly different terms, increasing property values for specific owners requires a general strategy of property redevelopment and change that relies on public rather than private action – a collective 'public' project of urban transformation. But tax increment financing means that the development of public property must be so arranged as to lead to a *private* good – rising property values. Such rising property values inevitably lead to the displacement of some property uses and some people, indeed, that is the whole intention of TIF. In many cities, this cycle of development leads to the destruction of inexpensive lodging for the poor (as noted in Chapter 4 for Minneapolis), the removal of social service agencies such as homeless shelters and soup kitchens that had located on previously marginal property, etc. (Dear and Wolch 1987; Mitchell and Staeheli, forthcoming). As poor people and unwanted uses (like cheap bars) are displaced – or actually replaced with 'higher and better uses' – then these people (the homeless and the formerly marginally housed, alcoholics) are forced onto public space: instead of drinking in cheap bars, poor alcoholics drink on the lawn of a park; instead of spending their days in a drop-in centre, homeless men and women hang out on the sidewalks of downtown. Yet the presence of the displaced in public space threatens the very rise in property values that the redevelopment of public space is meant to accomplish (Mitchell and Staeheli, forthcoming). Rules about the use of public space – who may be in it, what they may do – are often tightened, new forms of policing (such as business improvement district-sponsored private security forces) are developed, and a new, more exclusionary regime of public property develops. While the existence of a city as a conglomeration of privately and publicly held parcels – individual properties – may not require such a result, a city's status as landscape – a complete package – certainly does. Growth machine dynamics make such a result all but inevitable. Thus the challenge for the coalitions managing Sioux Falls' growth is to negotiate the set of contradictions posed by shaping an existing set of property uses and social relations into a form more conducive to the new economic interests that the city has recruited, while at the same time reinforcing the 'small town' virtuous landscape and 'we feeling' that is the very tool that was used to entice those new economies. We now turn to examine some of the contradictions that have resulted.

9.3 Property, landscape and marginalization in downtown Sioux Falls

In one sense, the project of redeveloping various properties of Sioux Falls, and at the same time remaking the landscape of the city, has been an unmitigated success. As the Development Foundation's website, clearly oriented towards attracting investors, points out.

> Since 1990, nearly $2.3 billion worth of construction has occurred in Sioux Falls. During the same period, 12,950 new housing units have been built and non-farm wage and salary employment has increased by 36,000 jobs. Companies have found Sioux Falls an ideal place to locate and expand operations.
>
> *(Original emphasis)*[13]

At the same time and despite all the new homes, homelessness has been steadily increasing in the city. A single-night count in August of 2001 by the Sioux Empire Homeless Coalition found a five-year high of 601 homeless in the city. Linking the continued growth of jobs in Sioux Falls with a relative economic decline elsewhere in the country, a local newspaper analysis suggested that 'People who come to Sioux Falls looking for jobs don't arrive with the resources to get housing' (Associated Press, 2001). This statement reveals the unevenness of the apparently smooth cycle of accumulation being produced in Sioux Falls, as well as the inevitability that such a rapid transformation in the landscape of the city has not been without its social costs.

In the battle over control of property, especially when property is enlisted in a highly successful and comprehensive renovation project such as that of Sioux Falls, those who do not own property have no stake in the process. Moreover, with the increase of value that is the intended consequence of redevelopment, the poor and propertyless (homeless, youth and low-wage renters) of Sioux Falls increasingly have no space in the city, as we will see. In spite of this, such groups are vitally important in considerations of the overall landscape renovation of the city. Though they are often required as workers for the production and reproduction of the city, they are at the same time a contradiction in the image of prosperity – the landscape vision – that city leaders are reliant upon to continue their development efforts. In Sioux Falls, specific conflicts over the use of downtown illustrate these points.

Phillips Avenue is the main artery of downtown commercialism and business. The street is lined with high-priced shops and bars and serves as the location of numerous summer festivals and street fairs (Figure 9.6). This current state of development belies the various forms Phillips Avenue has taken over the years. Even today, the area of prime entertainment venues is small, suggesting the strides city leaders have made only recently to enhance city attributes in accordance with the latest economic changes. While the

Figure 9.6 *Phillips Avenue shops*

focus of the economic development commencing with Citibank's entry to Sioux Falls has been on the northern boundaries of the city, the remake of downtown has also been an essential part of the new Sioux Falls.

According to a Downtown Sioux Falls employee, a 'big social change happened when Citibank came and brought upper management from New York. They were a whole different breed … You could see, you could feel the actual social changes in Sioux Falls that didn't go for the rest of the state'.[14] The social changes were, of course, necessary to the property development in Sioux Falls and created an impetus for city leaders to expand the reshaping of the city's landscape beyond the development parks. As a county welfare employee put it, city leaders asked themselves, 'how are we going to entertain them [those who moved in from big cities with the new businesses]?'[15] The answer was to establish 'high scale entertainment' including minor league professional sports teams with new stadiums, and the Washington Pavilion of Arts and Sciences with its world-class symphony hall.[16] As a series of discrete property developments, this sequential redevelopment seems quite non-contradictory – and indeed this is the picture of smoothly flowing growth that city leaders tout as the success story of Sioux Falls. However, the redevelopment of Sioux Falls was not only a series of discrete property changes – it was the reshaping of a whole landscape of social relations and lives that were often quite resistant to the new vision of Sioux Falls. Inevitably, the diversification of economic and social interests in Sioux Falls meant that some needs would not be met. In the class project of reshaping the city, the needs of capital accumulation take precedence, and so the people and ways of life that stand counter to these needs give shape to the contradictions Sioux Falls' leaders have to negotiate. They have had not only to develop property, but to remake the landscape *and* the social relations embodied in it. The competing uses of the downtown areas show who is benefiting from the remake and who is not.

Falls Park is a few blocks northeast of the prominent area of Phillips Avenue, and currently the area between the two draws considerably less investment than the Phillips Avenue shopping and entertainment district – and looks it. The street itself widens, overrunning the ample sidewalk space of the shopping area, and the businesses open are mainly junk dealers, scrap metal shops or other less permanent and less profitable establishments. The buildings are visibly more decayed, and more eclectically pieced together. The largest business on the north end of Phillips Avenue is the Pitts Inc. salvage yard, which has been in business there for 70 years. Walking from Phillips Avenue to Falls Park requires the negotiation of intermittent sidewalks and uneven verges along a frequently deserted route.

There is another way to get to Falls Park via downtown without driving a car. This route is along the bike path that accompanies the Big Sioux River through the city. The bike path is impeccably maintained through well-trimmed lawns with occasional benches. Especially on the way from downtown to Falls Park, the path winds through an almost continuous park past restaurant balconies and the city's beloved goose population. The bike path extends 14 miles (22 km.) around the city in a soon-to-be complete loop and is the source of much brochure-based boasting.

The idea for the Phillips to the Falls project is to extend the family and tourist-friendly atmosphere of the bike path to the streets between downtown and Falls Park. According to an article about the project in a South Dakota paper

> City officials hope a four-block extension of Phillips Avenue will transform a rough-hewn industrial area along the Big Sioux River into an urban showcase. The $4.5 million project aims to connect the downtown area more directly to the city's namesake and pre-eminent geographic feature: the falls of the Big Sioux River in Falls Park.
>
> (Aberdeen American News, 1997)

The city planning director said of the Phillips to the Falls project, 'It will make Falls Park much more a part of the city and give us the opportunity to clean up one of the biggest eyesores in the city' (Aberdeen American News, 1997). In May of 2001, the city overcame its last major hurdle towards the completion of the Phillips to the Falls project by purchasing the salvage yard from the Pitt's family after 13 years of negotiation. The completion of the project will allow for a smooth continuity in the revamped landscape of downtown Sioux Falls. As the city planner put it, 'Now when people think about our City, they think of the Park and the magnificent sight of the Sioux Falls rushing through the centre of our downtown area.'[17]

The project is part of an ongoing effort to revitalize Falls Park, which according to a member of the Development Foundation and the director of the Main Street Sioux Falls program (part of Downtown Sioux Falls), used to be full of 'drunks.'[18] Part of the process of cleaning up eyesores in the Park itself has been the removal of the 'drunks'. The director of the Main Street Sioux Falls program said that once a ban on alcohol had been passed and reinforced in Falls Park, the people who used to use the park had nowhere to go. The city, she said, would 'keep moving them around. We haven't solved the alcoholic problems, but Falls Park has undoubtedly increased its intake of visitors.'[19] There is now a laser light show every night at Falls Park that features a retelling of the founding of Sioux Falls and the importance of the Falls in the development of the city.

The revamping of the downtown-Falls Park corridor is part of Downtown Sioux Falls' efforts to remake the city for the consumers of 'world-class' culture, an endeavor which is highly similar to Auckland's goals, set out in Chapter 7. Perhaps the pinnacle of these efforts is the development of the Washington Pavilion of Arts and Sciences, a project approved by a narrow voter margin in 1993. The Washington Pavilion brings top performing and visual artists and touring acts to the city. But the very people the Pavilion seeks to attract – the relatively well-to-do management and upper-echelon workers in the outlying business parks, the suburban residents in the housing tracts – find the Pavilion's location in 'The Loop' a deterrent.

The Loop controversy, in which the Washington Pavilion stands at the symbolic centre, came to a head during the summer of 2000 in Sioux Falls. The debate raises issues of the changing allowable uses of the city streets, as well as the image of the city that is perceived beyond these uses. The Loop is a stretch of about four blocks that has been used by young people for cruising for generations. Recently, a man who had bought a house along the Loop, which is also being sold as part of the historic district of the city, began to complain of the noise from cars, stereos and youth occupying the Loop. A committee consisting of business leaders, city leaders and Loop users (teenagers, for the most part), was formed to investigate possible resolutions to the conflict. The committee came up with a set of proposals, including an anti-cruising ordinance, an anti-loitering ordinance, a curfew and the closing of a city parking lot dubbed 'Loser's Lot' by Loopers. The anti-cruising ordinance would allow police officers to arrest anyone passing by the same point more than three times in an hour and the closing of the lot plus the anti-loitering ordinance would restrict youth from congregating on the streets of downtown. Main Street Sioux Falls' director has taken an active part in the Loop controversy, advocating for the strictest possible laws to stop cruising. She said that though she did not consider the Loopers dangerous, their hanging out might be considered intimidating to Pavilion patrons, thereby threatening the nightlife being encouraged downtown. The controversy of the teenagers, who are repeating the behavior of generations past, versus the new downtown signals the tensions between a shifting social landscape and property development.

A member of the Downtown Sioux Falls committee described the 'battle to get the Pavilion', as a fight of 'Joe six-pack against the snooty-artsy, symphony people'.[20] But the Pavilion, or what it represents, is in contradiction to more than the recreation of the city's youth. As a part of a landscape that has been created for the specific aims of the class project of Sioux Falls' redevelopment, the Pavilion is equally a use of space and an assertion of class interests as a necessary part of redevelopment and growth. In order to perpetuate the Downtown Sioux Falls project, property value must be increased – thus the uses of downtown spaces that are not profitable must take a back seat to those that are. This, of course, means that those who cannot afford to contribute to the profit-making project increasingly have no space left. Beyond those who simply cannot afford to partake in the new amenities like the Washington Pavilion, perhaps like 'Joe six-pack', there are those like the Loopers or the homeless whose very presence runs counter to the mission of increasing property values.

'Joe six-pack' and the homeless might, however, still find some room at the north end of downtown – the traditional 'skid row' area of the city. Long 'a rough neighborhood, all seedy bars', the area serves as the home to the complex of buildings operating as shelter, drop-in centre and soup kitchen.[21] An initial building was donated by one of the city's car dealership owners to the Catholic diocese to start the centre. According to the program director, the centre was started despite the fact that the 'city fathers at a Rotary Club meeting said we don't have homeless people because everybody takes care of everybody'.[22] She claimed that the only reason city leaders allowed the centre to start downtown was because they thought it would be closed in a year and that the only reason people donated money for it was because of the influential people asking. Now, with the centre having been in operation for 15 years, the director is one of the most well known people in town and she realizes that her personal connections contribute to keeping the centre open and running. Without the specific people involved in its creation and operation, the centre would not continue to exist.

Even so, there is increasing pressure on the shelter and low-rent bars as gentrification has begun to take hold, with the presence of several new bars and clubs, including a Blues theme bar and a local brew-pub. The director of the homeless shelter indicated that since the new bars have begun to thrive, she has perceived 'subtle remarks in public meetings' concerning her organization's clients' presence on the streets around the bars.[23] She said she was glad when the new upscale bars replaced the seedy previous establishments, since they were often sites of fights or drunken loitering that she felt were a bad influence for her clients. Recently, a wrought iron fence was erected marking the barrier between the shelter and the renovated office building next door, supposedly to 'complement the neighborhood'.[24] The shelter director said, 'I remember the feeling when the fence went up, like a barrier. [The business man who owned the renovated office building] told the bishop [of the Catholic Diocese that owns the shelter buildings] that he didn't want loitering'.[25] In addition to the fencing in of the centre, the director has also faced explicit suggestions from a city planner that the centre should move out of downtown and toward the area of town near the meatpacking plant and the railroads where several other shelters are clustered. However, this suggestion seems to contradict the argument, made by people in the city planning office, that the city wants to avoid clustering people or city spaces by levels of wealth. The more likely explanation for the pressure to move the downtown shelter is that city planners are increasingly designating tourist locales and consumption centres, like downtown and Falls Park, as sites that exclude certain people. The result is that those who use the shelter, drop-in centre or eat at the soup kitchen are not only disenfranchised from city structures of wealth and ownership, but also from the very spaces of the city, including, perhaps especially, the city's putatively public spaces.

The development of the city has had clear advantages for some business owners and for those who have been able to secure jobs in the companies relocating to the Development Foundation's office parks, but the changes in the landscape of the city have not been without tensions. A city welfare worker said that when he was growing up in Sioux Falls in the 1960s divisions of wealth were visible in people's houses – the rich ones had basements and three floors, while the poor ones had only one floor. Now, he said, wealth disparities were much greater. Though the city still depends on low wage jobs, especially in construction and at the meatpacking plant, the ability of those who work at those jobs to secure a comfortable house and to otherwise become integrated into the city is decreasing. The class project that is Sioux Falls' property redevelopment is working to increase the divisions of class in the city, even as it constructs a landscape that is the very vision of prosperous success 'for all'.

9.4 Conclusion

Citibank is now the second largest employer in Sioux Falls, following Sioux Valley Hospital and Health Systems. Before Citibank, the John Morrell and Company pork producer was the city's largest employer. In wanting Citibank to move to Sioux Falls, the city was not just trying to recruit any jobs, but particular types of jobs. Sioux Falls' mayor at the time of Citibank's move, Rick Knobe, was quoted in *The New York Times* regarding the introduction of banking to the city: 'I can tell you one thing, everybody is talking about it. It's a good, clean industry' (New York Times, 1980). Citibank's arrival was not simply good fortune arising from a favorable arrangement of outside forces; the industry was courted by city officials who had a particular plan to change the image of Sioux Falls. The 'cleanness' of the banking industry has become inscribed in the landscape of the city – in the neat green lawns and smooth brick of the development parks that contrast so markedly from the dirt lots and ramshackle buildings of the meat processor.

Both to keep the new banking executives entertained, as popular rationale would have it, and to perpetuate the cycle of accumulation, the property redevelopment of Sioux Falls has not stopped with the business parks. The Development Foundation has overseen new institutions, like Downtown Sioux Falls, to manage the now nearly complete redevelopment of downtown properties like Phillips Avenue and Falls Park. The institutions have a dual purpose that well expresses their role as mediators of contradiction. Downtown Sioux Falls' primary purpose, and indeed its entire reason for existing, is to oversee the accumulation of profits from increased property value. But in doing this, Downtown Sioux Falls must also oversee the reshaping of a landscape that entails a variety of social uses that are in direct conflict to the interests of capital accumulation.

Beyond the spaces of the development parks, then, the various projects financed through rising property values have allowed an overhaul of much of the city from the downtown core to the residential fringes. This process has involved several social conflicts and the effective removal of people, like the Loopers or the homeless, whose practices and very existence have been viewed as barriers to accumulation. It has also created its own set of contradictions as different sections of the city and region have been pitted against one another in the competition for inward investment – for the capturing of the benefits of economic growth. Property development in capitalism is necessarily uneven, particularly so in Downtown. Indeed, Sioux Falls' history since the 1970s reveals that the efforts of Downtown Sioux Falls to renegotiate the economic and social landscapes of the city through property developments are not new. And these older landscapes – such as the mall – as well as new offshoots of the development parks, like new suburban strip malls, present competition for the downtown that further reinforces its singular drive to make the landscape as smooth as possible for profit. In the competition between various private property interests for profits within the city, downtown can simply not afford to make unprofitable concessions for contradictory social land uses. Turning urban space into a property growth machine will always create losers as well as winners. Growth is necessarily painful.

9.5 Notes

1. 'Tax increment financing is a system whereby property values in a particular district are frozen at a certain level; when property values rise, the taxes or the increased values are then funneled back into redevelopment projects there. TIFs are built on the concept that new value will be created, and that the future value can be used to finance part of the activities needed now to create new value' (Pepper, 1997).

2. The brewery is part of Falls Center, a collection of businesses that employ 139 professional and service industry employees and have increased the valuation of the National Historic Register building 841 per cent from $148,487, when it housed one business with less than 10 employees, to $1,397,423. 'The private investors were assisted by a collaboration of the City's historic preservation program, the federal Investment Tax Credit program (National Park Service, Department of the Interior), and the South Dakota State Property Tax Moratorium program' (*Seten*, 1999).

3. The Supreme Court of the USA has recognized the tension that exists between zoning and the rights of individual property owners. While it has upheld the constitutionality of zoning as a legitimate governmental function, it has also allowed individual property owners to sue for damages – or for exemption from regulation – if they feel their parcels have been inordinately affected by zoning (see Bruce, 1992).

4. Issues raised in the following discussion are developed more fully in Mitchell (1996, 2000, 2002, 2003). The direct link between landscape and property is made, in very different contexts, by Blomley (1998) and Cosgrove (1985, 1998). Excellent historical accounts of the tortured relationship between landscape, different juridical regimes of property and justice can be found in Olwig (1993, 1996).

5. Personal interview, July 2000.

6. Lincoln County, which became part of the Sioux Falls Metropolitan Statistical Area in 1992, was listed by USA Today as the second fastest growing US county.

7. Unemployment statistics fell from 4.6 per cent to 1.6 percent between 1980 and 1998 while population in the city increased by 44 per cent over the same time period, and 'ethnic population' as defined by the Sioux Falls Multicultural Center grew by 77 per cent between 1980 and 1990.

8. Sioux Falls was named 'best place to live in America' in 1992 and '#1 economy for 4th Consecutive year' in 1994 by *Money* magazine.

9. http://www.siouxfallsdevelopment.com/membership.cfm, accessed 16 April 2002.

10. How could it be otherwise, since property is developed a parcel at a time while the city as a whole has to be sold as a complete package.

11. Sioux Falls's Planning Department and the Development Foundation also carried out a 'community visioning project' that helped to solidify the overall growth project through the generation of goals and tasks related to the management of residential property interests with the growing diversity of the city (see Breitbach, 2001).

12. For reviews and analyses of the relationship between public space and urban redevelopment see Davis (1990); MacLeod (2002); McCann (1999); Mitchell (1997); Mitchell and Staeheli (forthcoming).

13. http://www.siouxfallsdevelopment.com/economic.cfm, accessed 19 April 2002.

14. Personal interview, July 2000.

15. Personal interview, July 2000.
16. Personal interview, July 2000.
17. http://www.epa.gov/region8/siouxfalls.html, accessed 20 April 2002.
18. Personal interview, July 2000.
19. Personal interview, July 2000.
20. Personal interview, July 2000.
21. Personal interview, July 2000.
22. Personal interview, July 2000.
23. Personal interview, July 2000.
24. Personal interview, July 2000.
25. Personal interview, July 2000.

Acknowledgements

We would like to thank Joseph Stoll, of Syracuse University for producing the map and Euan Hague, of De Paul University for his helpful comments.

Bibliography

ABERDEEN AMERICAN NEWS 1997: The Dakota's Sioux Falls plans downtown urban showcase. 28 October: A3.

ADAMS, D. 1994: *Urban Planning and the Development Process.* London: UCL Press.

ADAMS, J. S. and VAN DRASEK, B. 1993: *Minneapolis-St. Paul.* Minneapolis, MN: University of Minnesota Press.

AMBROSE, P. 1986: *Whatever Happened to Planning?* London: Methuen.

AN TAISCE 1985: Georgian Dublin: Policy for Survival. Dublin: Dublin City Association of An Taisce.

ARNSTEIN, S. 1969: A ladder of citizen participation. *Journal of the American Institute of Planners* 35, 216–24.

ASCHMAN, F. T. 1971: Nicollet Mall: civic cooperation to preserve Downtown's vitality. *Planners Notebook* 1 (6), 1–8.

ASCOT GROUP 2001: About us: Corporate Profile, http://www.theascot.com/abtus/body table.html#residence portfolio. Accessed April 23, 2001.

ASHTON, P. 1993: *The Accidental City.* Sydney: Hale and Iremonger.

ASSOCIATED PRESS, 2001: *Homeless number at five-year high.* The Associated Press State & Local Wire, 15 November. Available: http://www.lexis-nexis.com/universe.

AUSTIN, P. and WHITEHEAD, C. 1998: Auckland: cappuccino city? *Urban Policy and Research* 16, 233–40.

'AXISS AUSTRALIA' 2001: *The Prime Office Market in Australia, 2001.* Sydney: 'Axiss Australia'.

BACHRACH, P. and BARATZ, M. S. 1962: Power as non-decision-making. *American Political Science Review,* 947–52.

BALL, M. 1983: *Housing Policy and Economic Power: the Political Economy of Owner Occupation.* London: Methuen.

BALL, M. 1986: The built environment and the urban question. *Environment and Planning D: Society and Space* 4, 447–64.

BALL, M. 1994: The 1980s property boom. *Environment and Planning A,* 26, 671–95.

BALL, M. 1998: Institutions in British property research: a review. *Urban Studies* 9, 1501–17.

BALL, R. and PRATT, A. C. (eds) 1994: *Industrial Property: Policy and Economic Development.* London: Routledge.

BANNON, M. J., EUSTACE, J. G. and O'NEILL, M. 1981: *Urbanisation: Problems of Growth and Decay in Dublin.* National Economic and Social Council, Report 55. Dublin: Stationery Office.

BARRAS, R. 1979a: *The Returns from Office Development and Investment.* CES Research Series, 35. London: Centre for Environmental Studies.

BARRAS, R. 1979b: *The Office Property Cycle in London*. CES Research Series, 36. London: Centre for Environmental Studies.

BARRAS, R. and FERGUSON, D. 1987: Dynamic modeling of the building cycle: 2. Empirical results. *Environment and Planning A, 19, 4493–520.*

BARRETT, S. M., STEWART, M. and UNDERWOOD, J. 1978: *The Land Market and the Development Process: A Review of Research and Policy.* Occasional Paper no. 2, School for Advanced Urban Studies. Bristol: University of Bristol.

BASSETT, K. 1996: Partnerships, business élites and urban politics: new forms of urban governance in an English city? *Urban Studies* 33, 539–55.

BAYLEYS PROPERTY RESEARCH 1999: *Auckland Inner City Property Review*, http://www.bayleys.co.nz/publications/residentialApartJune99, accessed Nov. 2001.

BEAUREGARD, R. 1996: Between modernity and postmodernity: the ambiguous position of U.S. planning. In Campbell, S. and Fainstein, S. (eds), *Readings in Planning Theory*. Oxford: Blackwell, 213–34.

BELL, S. 1997: *Ungoverning the Economy: the Political Economy of Australian Economic Policy.* Melbourne: Oxford University Press.

BERG, L. VAN DEN, MEER, J. VAN DER and OTGAAR, A. H. J. 1999: *The Attractive City: Catalyst for Economic Development and Social Revitalisation*, Rotterdam: European Institute for Comparative Urban Research.

BERGER, J. 1972: *Ways of Seeing*. Harmondsworth: Penguin.

BERTZ, S. 2002: The peripheralisation of office development in the Dublin Metropolitan Area. *Irish Geography* 35 (2), 197–212.

BIRMINGHAM CITY COUNCIL 1988: *The Highbury Initiative*. Birmingham: BCC.

BIRMINGHAM CITY COUNCIL 2000: *Employment, Education and Skills in Birmingham.* Birmingham: BCC.

BIRMINGHAM CITY COUNCIL 2001: *The Beacon Council Scheme 2000, Application form on the theme of Town Centre Regeneration*. Birmingham: BCC.

BIRMINGHAM ECONOMIC INFORMATION CENTRE 2000: An insight into knowledge workers in Birmingham's labour market. *Labour Market Review* 2 (4), Spring/Summer, 2–3.

BLOMLEY, N. 1998: Landscapes of property. *Law and Society Review* 32, 567–612.

BLOMLEY, N. 2000: Property rights. In Johnston, R. (ed.) *The Dictionary of Human Geography*. Oxford: Blackwell (4th edition).

BODDY, M. 1981: The property sector in late capitalism: the case of Britain. In Dear, M. and Scott, A. J. (eds), *Urbanization and Urban Planning in Capitalist Society*. London: Methuen, 267–86.

BORG, N. 1973: Birmingham. In Holliday, J. (ed.) *City Centre Redevelopment: A study of British City Centre Planning and Case Studies of Five English City Centres*, London: Charles & Co, 30–77.

BOYLE, M. and HUGHES, G. 1994: The politics of urban entrepreneurialism in Glasgow. *Geoforum* 25, 453–70.

BREITBACH, C. 2001: *Urban Growth and Social Change in Sioux Falls, South Dakota.* Unpublished Master's Thesis, Geography Department. Syracuse, NY: Syracuse University.

BRENNER, N. 2001: The limits of scale? Methodological reflections on scalar structuration. *Progress in Human Geography* 25 (4), 591–614.

BRITTON, S., LE HERON, R. and PAWSON, E. 1992: *Changing Places in New Zealand: A Geography of Restructuring*. Christchurch: New Zealand Geographical Society.

BROADBENT, T. A. 1977: *Planning and Profit in the Urban Economy.* London: Methuen.

BROOKS, A. 1989: Commercial property: prairie growth; in Sioux Falls, commerce blooms like wildflowers. *New York Times*, 8 October, late edition: 10:19.

BROWNILL, S. 1990: *Developing London's Docklands: Another Great Planning Disaster?* London: Paul Chapman.

BRUCE, J. 1992: Zoning. In Hall, K. (ed.) *The Oxford Companion to the Supreme Court of the United States.* New York: Oxford University Press, 952–3.

BRYSON, J. R. 1997a: Business service firms, service space and the management of change. *Entrepreneurship and Regional Development* 9, 93–111.

BRYSON, J. R. 1997b: Obsolescence and the process of creative reconstruction, *Urban Studies* 34 (9), 1439–58.

BRYSON, J. R., DANIELS, P. W. and HENRY, N. 1996: From widgets to where? The Birmingham economy in the 1990s. In Gerrard, A. J. & Slater, T. R. (eds), *Managing a Conurbation: Birmingham and its Region.* Warwickshire: Brewin Books, 156–68.

BUDD, L. 1992: An urban narrative and the imperatives of the city. In Budd, L. & Whimster, S. (eds), *Global Finance and Urban Living: a Study of Metropolitan Change.* London: Routledge, 260–81.

BURKE, G. 1971: *Towns in the Making.* London: Edward Arnold.

BUTTLE WILSON 1987: *New Zealand Property Investment Companies.* New Zealand: Buttle Wilson Ltd.

CADMAN, D., AUSTIN-CROWE, L., TOPPING, R. and AVIS, M. 1991: *Property Development,* 3rd edition. London: E. & F. N. Spon.

CALUS, 1979 : *Buildings for Industry.* Reading: College of Estate Management, University of Reading.

CASINO CONTROL ACT 1990: *The Statutes of New Zealand*, 1990. Public Act 62, Wellington: Government Printer.

CASTELLS, M. 1978: *City, Class and Power.* London: Macmillan.

CHASE CORPORATION Ltd 1987: *Interim Report for the Twelve Months to 31 March 1987.* New Zealand: Chase Corporation.

CHASE CORPORATION Ltd 1989: *Annual Report.* New Zealand: Chase Corporation Ltd.

CHATWIN, J. 1999: The Brindleyplace masterplan. In Latham, I. and Swenarton, M. (eds), *Brindleyplace: a Model for Urban Regeneration.* London: Right Angle Publishing, 27–30.

CHERRY, G. E. 1994 *Birmingham: a Study in Geography, History and Planning.* Chichester: Wiley.

CITY OF MINNEAPOLIS PLANNING COMMISSION 1959: *Goals for Central Minneapolis: its function and design.* Minneapolis: City of Minneapolis.

CITY OF MINNEAPOLIS PLANNING DEPARTMENT 1978–1987: *State of the City.* Annual Report, Office of the City Co-ordinator. Minneapolis: City Planning Department.

CITY OF MINNEAPOLIS PLANNING DEPARTMENT 2000a: *Minneapolis Comprehensive Plan.* Minneapolis: Minneapolis Planning Department.

CITY OF MINNEAPOLIS PLANNING DEPARTMENT 2000b: *Downtown 2010.* Minneapolis: Minneapolis Planning Department.

CLARK, E. and LUND, A. 2000: Globalisation of a commercial property market: the case of Copenhagen. *Geoforum* 32, 467–75.

COAKLEY, J. 1994: The integration of property and financial markets. *Environment and Planning A,* 26, 697–713.

COBI *undated*: Communism: abolition of private property. *Proletarian, Journal of the Communist Organisation in the British Isles* 4, 14–54.

CONNELL, J. 2000: And the winner is ... In Connell, J. (ed.) *Sydney: The Emergence of a World City*. Melbourne: Oxford University Press, 1–18.

CONNOR, T. 1997: *Inner City Apartment Study, 1996*, Environmental Planning Division, City Planning Group. Auckland: Auckland City Council.

COSGROVE, D. 1985: Prospect, perspective, and the evolution of the landscape idea. *Transactions of the Institute of British Geographers* 10, 45–62.

COSGROVE, D. 1998: *Social Formation and Symbolic Landscape*. Madison: University of Wisconsin Press (2nd edition).

COX, K. R. 1973: *Conflict, Power and Politics in the City*. New York: McGraw Hill.

COX, K. R. 1981: Capitalism and conflict around the communal living space. In Dear, M. and Scott, A. J. (eds), *Urbanization and Urban Planning in Capitalist Society*. London: Methuen, 431–55.

COYLE, S. 1988: The Boston experience. In Blackwell, J. and Convery, F. (eds), *Revitalising Dublin: What Works?* Resource and Environmental Policy Centre. Dublin: University College Dublin, 99–105.

D'ARCY, É. and KEOGH, G. 1997: Towards a property market paradigm of urban change. *Environment and Planning A*, 29, 685–706.

DAHL, R. and LINDBLOM, C. 1953: *Politics, Economics and Welfare*. New York: Harper and Row.

DALY, M. 1982: *Sydney Boom, Sydney Bust: the City and its Property Market, 1850–981*. Sydney: Allen and Unwin.

DALY, M. 1987: Capital cities. In Jeans, D. N. (ed.) *Space and Society: Australia – a Geography, Volume 2*. Sydney: Sydney University Press, 75–111.

DALY, M. 1988: *Monitoring Sydney: Evaluating the Performance of the Sydney Metropolitan Region*. Australian Housing and Urban Research Institute Report. Brisbane: The University of Queensland Press.

DANIELS, P. W. and BRYSON, J. R. 2002: *The Professional Services Cluster in the City of Birmingham*. Birmingham: Birmingham Forward and Advantage West Midlands.

DAVIDOFF, P. 1965: Advocacy and pluralism in planning. *Journal of the American Institute of Planning* 31, 331–8.

DAVIES, J. G. 1972: *The Evangelistic Bureaucrat*. London: Tavistock.

DAVIS, J. and PRENDERGAST, T. 1995: Dublin. In Berry, J. and McGreal, S. (eds), *European Cities, Planning Systems and property Markets*. London: E & FN Spon, 193 216.

DAVIS, M. 1990: *City of Quartz: Excavating the Future in Los Angeles*. London: Verso.

DEAKIN, N. and EDWARDS, J. 1993: *The Enterprise Culture and the Inner City*. London: Routledge.

DEAR, M. 2000: *The Postmodern Urban Condition*. Oxford: Blackwell.

DEAR, M. and WOLCH, J. 1987: *Landscapes of Despair*. Princeton: Princeton University Press.

DEBENHAM, TEWSON and CHINNOCKS 1989 *Money Into Property*. London: Debenham, Tewson and Chinnocks.

DEHESH, A. and PUGH, C. 1999: The internationalisation of post-1980 property cycles and the Japanese 'bubble' economy, 1986–96. *International Journal of Urban and Regional Research* 23 (1), 147–64.

DEHESH, A. and PUGH, C. 2000: Property cycles in a global economy. *Urban studies*, 37, 13, 2581–602.

DEPARTMENT OF THE ENVIRONMENT AND LOCAL GOVERNMENT AND DEPARTMENT OF PUBLIC ENTERPRISE 2001: *New Institutional Arrangements for Land Use and Transport in the Greater Dublin Area*. Dublin: Department of the Environment and Local Government and Department of Public Enterprise.

DIXON, J. E., ERICKSEN, N. J., CRAWFORD, J. and BERKE, P. 1997: Planning under a co-operative mandate: new plans for New Zealand. *Journal of Environmental Planning and Management* 40 (5), 603–14.

DOMOSH, M. 1992: Corporate cultures and the modern landscape of New York City. In Anderson, K. and Gayle, F. (eds), *Inventing Places: Studies in Cultural Geography*. Melbourne: Longman, 72–85.

DRUDY, P. J. 1999: Dublin docklands: the way forward. In Killen, J. and MacLaran, A. (eds), *Dublin: Contemporary Trends and Issues for the Twenty-First Century*. Geographical Society of Ireland, Special Publication no. 11. Dublin: GSI, 35–47.

DUBLIN CORPORATION 1999: *Dublin City Development Plan*. Dublin: Dublin Corporation.

DUNCAN, J. 1990: *The City as Text: The Politics of Landscape Interpretation in the Kandyan Kingdom*. Cambridge: Cambridge University Press.

ELY, J. 1992: Property Rights. In Hall, K. (ed.) *The Oxford Companion to the Supreme Court of the United States*. New York: Oxford University Press, 683–91.

EWEN, L. A. 1978: *Corporate Power and Urban Crisis in Detroit*. Princeton, NJ: Princeton University Press.

FAGAN, R. 2000: Industrial change in the global city: Sydney's new spaces of production. In Connell, J. (ed.) *Sydney: The Emergence of a World City*. Oxford: Oxford University Press, 144–66.

FAINSTEIN, S. S. 1994: *The City Builders: Property, Politics and Planning in London and New York*. Oxford: Blackwell.

FAINSTEIN, S. S. 1995: *An Evaluation of the Minneapolis Neighborhood Revitalization Program*. New Brunswick, NJ: Center for Urban Policy Research, Rutgers University.

FAINSTEIN, S. S. 2000: New directions in planning theory. *Urban Affairs Review* 35 (4), 451–78.

FARRIER, D., LYSTER, K. and PEARSON, L. 1999: *Planning and Land use in New South Wales: the Environmental Law Handbook*. Sydney: Redfern Legal Centre Publishing.

FEAGIN, J. R. 1983: *The Urban Real Estate Game: Playing Monopoly With Real Money*. Englewood Cliffs, NJ: Prentice-Hall.

FINCHER, R. 1998: Planning for cities of difference. In Gleeson, B. and Hanley, P. (eds), *Renewing Australian Planning: New Challenges, New Agendas*. Canberra: Australian National University, 51–71.

FITCH, R. 1993 *The Assassination of New York*. London and New York: Verso.

FITZGERALD, S. 1992: *Sydney 1842–1992*. Sydney: Hale & Iremonger.

FOGELSONG, R. E. 1986: *Planning the Capitalist City: The Colonial Era to the 1920s*. Princeton, NJ: Princeton University Press.

FOTHERGILL, S., MONK, S. and PERRY, M. 1987: *Property and Industrial Development*. London: Hutchinson.

FRANKLIN, P. J. 1976: Insurance into property. *The Banker*, 126, 1127–9.

FREESTONE, R. 2000: Planning Sydney: historical trajectories and contemporary debates. In Connell, J. (ed.) *Sydney: The Emergence of a World City*. Melbourne: Oxford University Press, 119–43.

GAHAN, J. 1993: Fiscal measures in promoting urban regeneration: Custom House Docks, Dublin. In Berry, J., McGreal, S. and Deddis, W. (eds), *Urban Regeneration: Property Investment and Development*. London: E & FN Spon, 95–107.

GALASKIEWICZ, J. 1979: *Exchange Networks and Community politics*. Beverly Hills, CA: Sage Publications.

GALE, S. and MOORE, E. G. (eds) 1975: *The Manipulated City*. Chicago: Maaroufa Press.

GAYNOR, B. 1999: *Rookie Developer at Heart of Britomart*. New Zealand Herald, 20th March.

GIBSON, K. 1998: Social polarization and the politics of difference: discourses in collision of collusion. In Fincher, R. & Jacobs, J.M. (eds), *Cities of Difference*. London: Guildford, 301–16.

GILHOUSEN, M. 1985: *Shopping Centers and Retail Sales in the Twin Cities Metropolitan Area*, St Paul: Metropolitan Council of the Twin Cities Area.

GILHOUSEN, M. 1986: *Office Construction Update, 1983–1985, Twin Cities Metropolitan Area*. St Paul: Metropolitan Council of the Twin Cities Area.

GILHOUSEN, M. 1988, *Commercial Construction in the Twin Cities Metropolitan Area, 1987*. St Paul: Metropolitan Council of the Twin Cities Area.

GLEESON, B. J. 1995: The commodification of resource consent in New Zealand. *New Zealand Geographer* 51 (1), 42–8.

GLEESON, B. J. 1998: Commentary: globalisation and planning. *Environment and Planning A*, 30, 1143–7.

GLEESON, B. J. and GRUNDY, K. J. 1997: New Zealand's planning revolution five years on: a preliminary assessment. *Journal of Environmental Planning and Management* 40 (3), 293–313.

GLEESON, B. and LOW, N. 2000, *Australian Urban Planning: New Challenges, New Agendas*. Sydney: Allen and Unwin.

GLEESON, D. 2002: Dublin's north fringe: a new model for suburban greenfield development. *Journal of Irish Urban Studies* 1 (1), 81–5.

GOLDSMITH, M. 1980: *Politics, Planning and the City*. London: Hutchinson.

GOODCHILD, B. 1990: Planning and the modernism/post-modernism debate. *Town Planning Review* 61, 119–37.

GOODCHILD, R. N. 1978: The operation of the private land market. In Pearce, B. J., Curry, N. R. and Goodchild, R. N. (eds), *Land Planning and the Market* Department of Land Economy, Occasional Paper no. 9. Cambridge: University of Cambridge, 11–47.

GOODCHILD, R. and MUNTON, R. 1985: *Development and the Landowner*. London: George Allen & Unwin.

GREEN, S. 1986: *Who Owns London?* London: Weidenfeld and Nicholson.

GREER, S. and ORLEANS, P. 1962: The mass society and the parapolitical structure. *American Sociological Review* 27, 634–46.

GRUNDY, K. J. and GLEESON, B. J. 1996: Sustainable management and the market. *Land Use Policy* 13 (3), 197–211.

GUNDER, M. 2000: Urban policy formation under efficiency: the case of the Suckland City Council's Britomart project. In Memon, A. and Perkins, H. (eds), *Environmental Planning and Management in New Zealand*. New Zealand: Dunmore Press, 294–308.

GUY, S. and HENNEBERRY, J. 2000: Understanding urban development processes: integrating the economic and the social in property research. *Urban Studies* 37, 13, 2399–416.

HAMMERSON PLC 1999: *The Birmingham Alliance.* Press release issues 25th February, 1999. London: Hammerson.

HAMNETT, G. 2000: The late 1990s: competitive versus sustainable cities. In Freestone, R. & Hamnett, G. (eds), *The Australian Metropolis: a Planning history.* Sydney: Allen & Unwin, 168–89.

HANSON, R. and McNAMARA, J. 1981: *Partners.* Minneapolis: The Dayton Hudson Foundation.

HARVEY, D. 1977: Labor, capital, and class struggle around the built environment in advanced capitalist societies. *Politics and Society* 6, 265–95.

HARVEY, D. 1978: Urbanization under capitalism: a framework for analysis. *International Journal of Urban and Regional Research* 2, 101–31.

HARVEY, D. 1981: The urban process under capitalism: a framework for analysis. In Dear, M. and Scott, A. J. (eds), *Urbanization and Urban Planning in Capitalist Society.* London: Methuen, 91–121.

HARVEY, D. 1982: *The Limits to Capital.* Oxford: Blackwell; Chicago: University of Chicago Press.

HARVEY, D. 1985: *The Urbanization of Capital: Studies in the History and Theory of Capitalist Urbanization.* Baltimore, MD: John Hopkins University Press.

HARVEY, D. 1989: *The Urban Experience.* Oxford: Blackwell.

HARVEY, D. 2000: *Spaces of Hope.* Edinburgh: Edinburgh University Press.

HEALEY, P. 1991: Models of the development process: a review. *Journal of Property Research* 8, 219–38.

HEALEY, P. 1992: An institutional model of the development process. *Journal of Property Research* 9, 33–44.

HEALEY, P. 1994: Urban policy and property development: the institutional relations of real-estate development in an old industrial region. *Environment and Planning A,* 26, 177–98.

HEALEY, P. 1996: Planning through debate: the communicative turn in planning theory. In Campbell, S. and Fainstein, S. (eds), *Readings in Planning Theory.* Oxford: Blackwell, 234–58.

HEALEY, P. 1997: *Collaborative Planning: Shaping Places in Fragmented Societies.* London: Macmillan.

HEALEY, P. 1998: Regulating property developments and the capacity of the development industry. *Journal of Property Research,* 15 (3), 211–27.

HEALEY, P. and BARRETT, S. M. 1990: Structure and agency in land and property development processes: some ideas for research. *Urban Studies,* 27 (1) 89–104.

HEALEY, P., CAMERON, S., DAVOUDI, S., GRAHAM, S. and MADANI, A. (eds) 1995: *Managing Cities: The New Urban Context.* Chichester: Wiley.

HEALEY, P., DAVOUDI, S., TAVSANOGLU, S., O'TOOLE, M. and USHER, D. (eds) 1992: *Rebuilding the City: Property-led Urban Regeneration.* London: E. & F. N. Spon.

HEATH, R. 1984: *History of Downtown Planning.* Address to Metro 2000 Committee, 1984, mimeo May 20, 1985.

HENNEBERRY, J. 1994: High technology firms and the property market. In Ball, R. and Pratt, A. (eds), *Industrial Property: Policy and Economic Development.* London: Routledge, 106–28.

HILLIER, J. and SEARLE, G. 1995: *Rien Ne Va Plus: Fast Track Development and Public Participation in Pyrmont-Ultimo, Sydney.* Sydney Vision: UTS Papers in Planning No. 3,

Planning Program, Faculty of Design, Architecture and Building. Sydney: University of Technology.

HIORNS, F. R. 1956: *Town Building in History*. London: G. G. Harrap.

HOLLOWAY, J. and PICCIOTTO, S. 1978: Introduction: Towards a materialist theory of the state. In Holloway, J. and Picciotto, S. (eds), *State and Capital: a Marxist Debate*. London: Edward Arnold, 1–31.

HOUGHTON EVANS, W. 1978: *Planning Cities: Legacy and Portent*. London: Lawrence and Wishart.

IMRIE, R. and THOMAS, H. 1995: Urban policy processes and the politics of urban regeneration. *International Journal of Urban and Regional Research* 19, 479–94.

IMRIE, R. and THOMAS, H. (eds) 1999: British urban policy: an evaluation of the urban development corporations. London: Sage Publications.

INVESTMENT PROPERTY DATABANK 1991: *IPD Property Investors Digest, 1991*. London: Investment Property Databank.

IRISH TIMES, 2002: B of I headquarters fetches €82.5m. *Irish Times*, April 4th.

JOHNSON, R. 2002: Britain's boom town. *The Sunday Times Magazine*, July 28th: 46–55.

JOHNSTON, R. J. 1979: *Political, Electoral and Spatial Systems*. Oxford: Oxford University Press.

JOHNSTON, R. J., GREGORY, D. and SMITH, D. 1994: *The Dictionary of Human Geography*. Oxford: Blackwell, 3rd edition.

JONAS, A. and WILSON, D. (eds) 1999: *The Urban Growth Machine: Critical Perspectives Two Decades Later*. Albany: State University of New York Press.

KELSEY, J. 1995: *The New Zealand Experiment: A World Model for Structural Adjustment?* Auckland: Auckland University Press/Bridget Williams Books.

KEOGH, G. and D'ARCY, É. 1999: Property market efficiency: an institutional economics perspective. *Urban Studies* 36 (13), 2401–14.

KEY, T., ESPINET, M. and WRIGHT, C. 1990: Prospects for the property industry: an overview. In Healey, P. and Nabarro, R. (eds), *Land and Property Development in a Changing Context*. Aldershot: Gower, 17–44.

KIRK, G. 1980: *Urban Planning in a Capitalist Society*. London: Croom Helm.

KIVELL, P. 1993: *Land and the City: Patterns and Processes of Urban Change*. London: Routledge.

KNEVITT, C. 1985: *Space on Earth: Architecture: People and Buildings*. London: Thames Methuen.

KNOX, P. L. 1982: *Urban Social Geography: An Introduction*. London: Longman.

KNOX, P. L. 1987: *Urban Social Geography: An Introduction*. Second Edition. Harlow: Longman.

KNOX, P. L. 1993: Capital, material culture, and socio-spatial differentiation. In Knox, P. (ed.) *The Restless Urban Landscape*. Englewood Cliffs, NJ: Prentice Hall, 1–34.

KNOX, P. L. (ed.) 1993: *The Restless Urban Landscape*. Englewood Cliffs, NJ: Prentice Hall.

KNOX, P. and CULLEN, J. 1981: Planners as urban managers: an exploration of the attitudes and self-image of senior British planners. *Environment and Planning* A, 13, 885–98.

KRUMHOLZ, N. 1994: Dilemmas of equity planning: a personal memoir *Planning Theory*, 10/11, 45–58.

LAMARCHE, F. 1976: Property development or the economic foundations of the urban question. In Pickvance, C. (ed.) *Urban Sociology*. London: Methuen, 85–118.

LATHAM, I. and SWENARTON, M. (eds) 1999: *Brindleyplace: a Model for Urban Regeneration*. London: Right Angle Publishing.

LEES, L. and BERG, L. 1995: Ponga, glass and concrete: a vision for urban socio-cultural geography in Aotearoa/New Zealand. *New Zealand Geographer* 51 (2), 32–41.

LEITNER, H. 1990: Cities in pursuit of economic growth: the local state as entrepreneur. *Political Geography Quarterly* 9 (2), 147–70.

LEITNER, H. 1994: Capital markets, the development industry, and urban office market dynamics: rethinking building cycles. *Environment and Planning A*, 26, 779–802.

LEITNER, H. and GARNER, M. 1993: The limits of local initiatives: a reassessment of urban entrepreneurialism for urban development. *Urban Geography*, 14, 57–77.

LEY, D. 1996: *The New Middle Class and the Remaking of the Central City*. Oxford: Oxford University Press.

LOFTMAN, P. and NEVIN, B. 1996: Prestige urban regeneration projects: socio-economic impacts. In Gerrard, A. J. and Slater, T. R. (eds), *Managing a Conurbation: Birmingham and its Region*. Warwickshire: Brewin Books, 187–97.

LOGAN, J. 1993: Cycles and trends in the globalization of real estate. In Knox, P. L. (ed.) *The Restless Urban Landscape*. Englewood Cliffs, NJ: Prentice Hall, 35–54.

LOGAN, J. R. and MOLOTCH, H. L. 1987: *Urban Fortunes: the Political Economy of Space*. Berkeley, CA : University of California Press.

LOW, S. 2000: *On the Plaza: The Politics of Public Space and Culture*. Austin: University of Texas Press.

LUITHLEN, L. 1992: Marxian concepts, capital accumulation and office property. *Journal of Property Research* 9, 227–46.

LUITHLEN, L. 1993: Capital accumulation and office development in Leicester, 1967–90. *Journal of Property Research* 10, 27–48.

MacLARAN, A. 1993: *Dublin: the Shaping of a Capital*. London: Belhaven-Wiley.

MacLARAN, A. 1996: *Office development in Dublin and the tax incentive areas*. Irish Geography, (2), 1996.

MacLARAN, A. and BEAMISH, C. 1985: Industrial property development in Dublin, 1960–1982. *Irish Geography* 18, 37–50.

MacLARAN, A. and KILLEN, J. 2002: The suburbanisation of office development in Dublin and its transport implications. *Journal of Irish Urban Studies* 1 (1), 21–35.

MacLARAN, A., MacLARAN, M. and MALONE, P. 1987: Property cycles in Dublin: the anatomy of boom and slump in the industrial and office property sectors. *Economic and Social Review* 18 (4), 237–56.

MacLARAN, A. and O'CONNELL, R. 2001: The changing geography of office development in Dublin. In Drudy, P. J. & MacLaran, A. (eds), *Dublin: Economic and Social Trends, Vol. 3*. Centre for Urban and Regional Studies. Dublin: Trinity College, 25–37.

MacLARAN, A., WILLIAMS, B. and EMERSON, H. 1995: *Residential Development in Central Dublin: a Survey of Current Occupiers*. Centre for Urban & Regional Studies. Dublin: Trinity College.

MacLEOD, G. 2002: From urban entrepreneurialism to a 'revanchist city'? On the spatial injustices of Glasgow's renaissance. *Antipode* 34, July, 602–24.

MADELIN, R. 1999a: Planning for the Market. In Latham, I. & Swenarton, M. (eds) *Brindleyplace: a Model for Urban Regeneration*. London: Right Angle Publishing: 32–4.

MADELIN, R. 1999b: Lessons from Brindleyplace. In Latham, I. & Swenarton, M. (eds) *Brindleyplace: a Model for Urban Regeneration*. London: Right Angle Publishing, 46–7.

MALONE, P. 1985: *Office Development in Dublin 1960–1983: Property, Profit and Space.* Unpublished Ph.D. Thesis, Department of Geography. Dublin: Trinity College Dublin.

MALONE, P. 1996: Dublin: motive, image and reality in the Custom House Docks, in Malone, P. (ed.) *City, Capital and Water.* London: Routledge, 65–89.

MARRIOT, O. 1967: *The Property Boom.* London: Hamish Hamilton/Pan.

MARTIN, R. and MINNS, R. 1995: Undermining the financial basis of regions: the spatial structure and implications of the UK pension fund system. *Regional Studies* 29 (2), 125–44.

MASSEY, D. and CATALANO, A. 1978; *Capital and Land: Landownership by Capital in Great Britain.* London: Edward Arnold.

McCANN, E. 1999: Race, protest and place: contextualizing Lefebvre in the US city. *Antipode* 31, 163–84.

McCARTHY, J. 2002: Entertainment-led regeneration: the case of Detroit. *Cities* 19 (2), 105–11.

McDONALD, F. 2000: *The Construction of Dublin.* Dublin: Gandon Editions.

McDONALD, L. 2001: Commentary: Residents Paying for Minneapolis Heavy Use of TIF. *The Star Tribune* (6/10/01).

McGUIRK, P. M. 1991: *Perspectives on the Nature and Role of Urban Planning in Dublin.* Unpublished Ph.D. thesis, Department of Geography. Dublin: Trinity College.

McGUIRK, P. M. 1994: Economic restructuring and the realignment of the urban planning system: the case of Dublin. *Urban Studies* 31 (2), 1994, 289–307.

McGUIRK, P. M. 1995: Power and influence in urban planning: community and property interests' participation in Dublin's planning system. *Irish Geography* 28, 64–75.

McGUIRK, P. M. 2001: Situation communicative planning theory: context, power, and knowledge. *Environment and Planning* A, 33, 195–217.

McGUIRK, P. M. and MacLARAN, A. 2001: Changing approaches to urban planning in an 'entrepreneurial city': the case of Dublin. *European Planning Studies* 9 (4), 2001, 437–57.

McGUIRK, P. and O'NEILL, P. 2002: Planning a prosperous Sydney: the challenges of planning urban development in the new urban context. *Australian Geographer* 33, in press.

McINERNEY, J. 1998: Local government planning and legislative reform. *Australian Planner* 35, 143–6.

McKAY, D. M. and COX, A. W. 1979: *The Politics of Urban Change.* London: Croom Helm.

MEINIG, D. (ed.) 1979: *The Interpretation of Ordinary Landscapes.* New York: Oxford University Press.

MEMON, A. 1991: Shaking off a colonial legacy? Town and country planning in New Zealand, 1870s–1980s. *Planning Perspectives* 16, 19–32.

METROPOLITAN COUNCIL 1971: *Metropolitan Development Guide – Major Diversified Centers – Policies, System Plan, Program.* St Paul: Metropolitan Council.

MERWIN, J. 1983: Citibank put little South Dakota back on the map. *Forbes,* 21st November, 85–92.

MEYERS, M. 2001: The $62 Million Target. *The Star Tribune* (6/10/01).

MILROY-MOORE, B. 1996: Some thoughts about difference and pluralism. In Campbell, S. and Fainstein, S. (eds), *Readings in Planning Theory.* Oxford: Blackwell, 461–6.

MINNEAPOLIS COMMUNITY DEVELOPMENT AGENCY (MCDA) 1987a: *City of Minneapolis Tax Increment Districts: Summary Report.* Minneapolis: M.C.D.A.

MINNEAPOLIS COMMUNITY DEVELOPMENT AGENCY (MCDA) 1987b: *Tax Increment Financing Issues,* Minneapolis: M.C.D.A.

MINNEAPOLIS COMMUNITY DEVELOPMENT AGENCY (MCDA) 1988: *Personal communication.*

MINNEAPOLIS COMMUNITY DEVELOPMENT *Web Site,* 2001: *www. mcda.org.* Subject search of downtown development issues (7/01/01).

MINNEAPOLIS NEIGHBORHOOD REVITALIZATION PROGRAM (NRP) 2000: *1990–2000 Progress Report.* Minneapolis: NRP.

MINNEAPOLIS PLANNING AND DEVELOPMENT 1970: *Metro Center '85.* Minneapolis: Minneapolis Planning Department.

MINNEAPOLIS PLANNING DEPARMENT *Web Site.* 2001: www. *ci. minneapolis.mn.us.* Subject search of city planning issues (7/01/01).

MITCHELL, D. 1996: *The Lie of the Land: Migrant Workers and the California Landscape.* Minneapolis: University of Minnesota Press.

MITCHELL, D. 1997. The annihilation of space by law: the roots and implications of anti-homeless laws in the United States. *Antipode* 29, 303–35.

MITCHELL, D. 2000: *Cultural Geography: A Critical Introduction.* Oxford: Blackwell.

MITCHELL, D. *forthcoming* 2003: California living, California dying: dead labor and the political economy of landscape. In Anderson, K., Pile, S. and Thrift, N. (eds), *Handbook of Cultural Geography.* London: Sage.

MITCHELL, D. *forthcoming* 2003: Landscape. In Sibley, D., Atkinson D., Jackson, P. and Washbourne, N. (eds), *Critical Concepts in Cultural Geography.* London: I. B. Taurus.

MITCHELL, D. and STAEHELI, L. *forthcoming* 2003: Clean and safe? Property redevelopment, public space and homelessness in downtown San Diego. In Low, S. and Smith, N. (eds), *The Politics of Public Space.*

MOLOTCH, H. 1976: The city as growth machine: toward a political economy of place. *American Journal of Sociology* 82, 309–30.

MONTGOMERY, J. 1995: The story of Temple bar: creating Dublin's cultural quarter. *Planning Practice and Research* 10, 101–10.

MORGAN, G. 1991: History on the Rocks. In Rickard, J. and Spearritt, P. (eds), *Packaging the Past: Public Histories.* Melbourne: Melbourne University Press, 78–87.

MORICZ, Z. 1994: *Shaping the Built Environment: the Creation of Office Space in Auckland City During 1975–1990.* Unpublished Masters Thesis, Department of Geography. Auckland: University of Auckland.

MORICZ, Z. and MURPHY, L. 1997: Space traders: reregulation, property companies and Auckland's office market, 1975–1994. *International Journal of Urban and Regional Research* 21 (2), 165–79.

MORRIS, A. E. J. 1972: *History of Urban Form.* London: Godwin.

MUMFORD, L. 1961: *The City in History.* London: Secker and Warburg.

MURPHY, L. 1996: Gambling on casinos. In Le Heron, R. and Pawson, E. (eds), *Changing Places: New Zealand in the Nineties.* Auckland: Longman Paul, 343–4.

MURPHY, L., KEARNS, R. A. and FRIESEN, W. 1999: Transforming the city: people, property and identity in millennial Auckland. *New Zealand Geographer* 55(2), 60–5.

NABARRO, R. 1990: The investment market in commercial and industrial development: some recent trends. In Healey, P. & Nabarro, R. (eds), *Building the New Britain: Land and Property Development Processes in a Changing Context.* Aldershot: Gower, 47–59.

NABARRO, R. and KEY, T. 1992: Current trends is commercial property investment and development. In Healey, P., Davoudi, S., O'Toole, M., Tavsanoglu, S. and Usher, D. (eds), *Rebuilding the City: Property-led Urban Regeneration*. London: Spon, 45–59.

NATIONAL BUSINESS REVIEW 1994: *Chase: How Convoluted Accounting Concealed Unsound Financing*. National Business Review, 34–5.

NEW YORK TIMES 1980: Highlights South Dakota: the birth of a banking center. *New York Times*, 29th June, 3: 17.

NEWLAND, O. 1994: *Lost Property: The Crash of '87 ... and the Aftershock*. Auckland: Harper Collins.

NEWMAN, P. and THORNLEY, A. 1996: *Urban Planning in Europe*. London: Routledge.

NEWMAN, P. and THORNLEY, A. 1997: Fragmentation and centralisation in the governance of London: influencing the urban policy and planning agenda. *Urban Studies* 34, 967–88.

NSW DEPARTMENT OF PLANNING 1994: *Affordable Housing: Draft City West Affordable Housing Program*; Draft Sydney Regional Environmental Plan No. 26, Amendment No. 4. Sydney: NSW Department of Planning.

O'CONNOR, K., STIMSON, R. and DALY, M. 2001: *Australia's Changing Economic Geography: a Society Dividing*. Meridian Series. Melbourne: Oxford University Press.

O'DONNELL, J. 1989: The entrepreneurial developer. *Urban Land* 48, 7, 34–5.

O'NEILL, P. and McGUIRK, P. 2002: Prosperity along Australia's eastern seaboard: Sydney and the geopolitics of urban and economic change. *Australian Geographer* 33, 241–61.

OFFICE OF THE DEPUTY PRIME MINISTER 2000: *Training for Urban Design*. London: Cabinet Office.

OFFICE OF THE LEGISLATIVE AUDITOR, STATE OF MINNESOTA, PROGRAM EVALUATION DIVISION 1986: *Tax Increment Financing*. St Paul: State of Minnesota.

OLWIG, K. 1993: Sexual cosmology: nation and landscape at the conceptual interstices of nature and culture: or what does landscape really mean? In Bender, B. (ed.) *Landscape: Politics and Perspectives*. Oxford: Berg, 307–43.

OLWIG, K. 1996: Recovering the substantive nature of landscape. *Annals of the Association of American Geographers* 86, 630–53.

ORCHARD, L. 1995: National Urban Policy in the 1990s. In Troy, P. (ed.) *Australian Cities: Issues, Strategies and Policies for Urban Australia in the 1990s*. Melbourne: Cambridge University Press, 65–86.

ORFIELD, M. 1997: *Metropolitics*. Washington, DC: Brookings Institution.

ORGANIZING COMMITTEE OF THE CITY OF ST PAUL CENTRAL BUSINESS DISTRICT AUTHORITY 1943: *Downtown St Paul: a Plan for its Development*. St Paul: St Paul C.B.D. Authority.

OSBORN, F. and WHITTICK, A. 1977: *New Towns: their Origins, Achievements and Progress*. London: Leonard Hill.

PAINTER, J. 1997: Regulation, regime and practice in urban politics. In Lauria, M. (ed.) *Reconstructing Urban regime Theory: Regulating Urban Politics in a Global Economy*. Thousand Oaks, CA, Sage, 122–43.

PAINTER, M. 1997: Reshaping the public sector. In Galligan, B., McAllister, I. and Ravenhill, J. (eds), *New Developments in Australian Politics*. Melbourne: Macmillan Education, 148–55.

PARTRIDGE, D. 1999: Buildings and Squares. In Latham, I. & Swenarton, M. (eds), *Brindleyplace: a Model for Urban Regeneration*. London: Right Angle Publishing, 49–53.

PASHUKANIS, E. (1924) [1978]: *Law and Marxism: a General Theory*. London: Ink Links.

PAWSON, E. 1996: The state and social policy. In Le Heron, R. and Pawson, E. (eds), *Changing Places: New Zealand in the Nineties*. Auckland: Longman Paul, 210–46.

PAWSON, E. 1999: Remaking places. In Le Heron, R., Murphy, L., Forer P. and Goldstone M. (eds), *Explorations in Human Geography: Encountering Place*. Auckland: Oxford University Press, 346–66.

PECK, J. 1995: Moving and shaking: business élites, state localism and urban privatism. *Progress in Human Geography* 19, 16–46.

PECK, J. and TICKELL, A. 1994: Searching for a new institutional fix. In Amin, A. (ed.) *Post-Fordism: a Reader*. London: Blackwell, 280–316.

PEEL, M. 1995: The urban debate: from 'Los Angeles' to the urban village, in Troy, P. (ed.) *Australian Cities: Issues, Strategies and Policies for Urban Australia in the 1990s*. Melbourne: Cambridge University Press, 39–65.

PEISER, R. 1990: Who plans America? Planners or developers? *American Planning Association Journal* 56, 496–503.

PEPPER, E. 1997: *Lessons from the Field: Unlocking Development Potential with an Environmental Key*. Northeast Midwest Institute. Retrieved April 8, 2002, from Northeast Midwest Institute Web site: http://www.nemw.org/lessons.htm.

PRATT, A. and BALL, R. 1994: Industrial property, policy and economic development. In Ball, R. & Pratt, A. (eds), *Industrial Property: Policy and Economic Development*. London: Routledge, 1–19.

PRYKE, M. 1994a: Looking back on the space of a boom: (re)developing spatial matrices in the City of London. *Environment and Planning A*, 26, 235–64.

PRYKE, M. 1994b: Urbanizing capitals: towards an integration of time, space and economic calculation. In Corbridge, S., Thrift, N. and Martin, R. (eds), *Money, Power and Space*. Oxford: Blackwell, 218–52.

PUBLIC FINANCIAL SYSTEMS 1987: *Phase I Benefit Analysis for the New Nicollet Mall*. Minneapolis: Public Financial Systems.

RATCLIFFE, J. and STUBBS, M. 1996: *Urban Planning and Real Estate Development*. London: UCL Press.

REID, M. 1982: *The Secondary Banking Crisis, 1973–75*. London: Macmillan.

RESOURCE MANAGEMENT ACT 1991. *The Statutes of New Zealand*, 1991. Public Act 69, Wellington: Government Printer.

REUSS, H. S. 1977: *To Save Our Cities: What Needs to be Done*. Washington, DC: Public Affairs Press.

RHODES, R. 1988: *Beyond Whitehall and Westminster: the Sub-Central Government of Britain*. London: Unwin Hyman.

ROWEIS, S. T and SCOTT, A. J. 1981: The urban land question. In Dear, M. and Scott, A. J. (eds), *Urbanization and Urban Planning in Capitalist Society*, London: Methuen, 123–57.

ROYAL TOWN PLANNING INSTITUTE, 2001: *Special Award for Acclaimed Birmingham Planners*. Press Release No PR2001/24 Issued 7th February 2001, London: RTPI.

SANDERCOCK, L., 1998: *Towards Cosmopolis*. Chichester: John Wiley.

SAUER, C. 1925 [1963]: The morphology of landscape. In Leighly, J. (ed.) *Land and Life: A Selection of the Writings of Carl Ortwin Sauer*. Berkeley: University of California Press, 315–50.

SAUNDERS, P. 1979: *Urban politics: A sociological interpretation*. London: Hutchinson.

SCHAFFER, F. 1972: *The New Town Story*. London: Paladin.

SCHWARTZ, A. and GLICKMAN, N. 2000: *Rebuilding Downtown: A Case Study of Minneapolis*. Draft report to the National Center for Revitalization of Central Cities. Center for Urban Policy Research. New Brunswick, NJ: Rutgers University.

SCOTT, A. J. 1980: *The Urban Land Nexus and the State*. London: Pion.

SCOTT, A. J. 2002: *Global City Regions: Trends, Theory, Policy*. Oxford: Oxford University Press.

SCOTT, A. J. and ROWEIS, S. T. 1977: Urban planning in theory and practice: a reappraisal. *Environment and Planning* A, 9, 1097–119.

SEARLE, G. 1998: Regime, regulation, and two decades of central Sydney development. In Freestone, R. (ed.) *The Twentieth Century Urban Planning Experience*. Proceedings of the 8th International Planning Society Conference and 4th Australian Planning/Urban History Conference, 15–18 July. Sydney: University of New South Wales, 805–10.

SEARLE, G. 2001: *Urban Planning as an Instrument of State Corporatism*. Paper presented to the World Planning Schools Congress, Shanghai, July 11–15.

SEARLE, G. and BOUNDS, M. 1999: State powers, state land and competition for global entertainment: the case of Sydney. *International Journal of Urban and Regional Research* 23, 165–72.

SETEN, D. 1999: Dialogue no. 31 – Seten Presentation Materials. Retrieved April 8, 2002, from Institute for Responsible Management website: http://www.instrm.org/dialogue/dialog31/seten/seten31.htm

SIMPSON, W. 1994: *The Administration and Implementation of Strategic Planning and Development Control Within the Sydney Region: Report to the Honourable Robert Webster, Minister for Planning and Minister for Housing* (Commissioner William Simpson, Chairman). Sydney: Office of the Commissioners of Inquiry for Environment and Planning.

SLATER, T. R. 1996: Birmingham's Black and South-Asian population'. In Gerrard, A. J. and Slater, T. R. (eds), *Managing a Conurbation: Birmingham and its Region*. Warwickshire: Brewin Books; 140–55.

SMITH, N. 1996: *The New Urban Frontier: Gentrification and the Revanchist City*. London: Routledge.

SMYTH, H. 1982: *Land Banking, Land Availability and Planning for Private House Building*, School for Advanced Urban Studies, Working Paper 23. Bristol: University of Bristol.

SMYTH, H. 1985: *Property Companies and the Construction Industry in Britain*. Cambridge: Cambridge University Press.

SMYTH, H. 1994: *Marketing the City: the Role of Flagship Developments in Urban Regeneration*. London: E. & F. N. Spon.

STOKER, G. and YOUNG, J. 1993: *Cities in the 1990s*. London: Longman.

STORPER, M. 1998: *The Regional World*. New York: The Guilford Press.

SUTCLIFFE, A. 1981: *Towards the Planned City: Germany, Britain, the United States and France, 1780–1914*. Oxford: Blackwell.

SYDNEY CITY COUNCIL 1997: *1997 Floorspace and Employment Survey, Sydney Central City Area Summary Results*. Sydney: Sydney City Council.

SYDNEY CITY COUNCIL 1999: *Economic Development and Viability of the Central City: A Discussion Paper from the Council of the City of Sydney*. Information Sharing Paper. Sydney: Sydney City Council.

SYDNEY CITY COUNCIL 2000: *Sydney 1999: City of Sydney Yearbook*. Sydney: Sydney City Council.

SYDNEY CITY COUNCIL 2001: *City Commercial Monitor*, December, Issue 4. Sydney: Sydney City Council.

SYDNEY CITY COUNCIL 2002: *City Residential Monitor*, June, Issue 26. Sydney: Sydney City Council.

TALEN, E. 1999: Sense of community and neighbourhood form: an assessment of the social doctrine of new urbanism. *Urban Studies* 36, 1361–79.

TAYLOR, M. 2000: *The Regeneration of Birmingham City Centre – 1985–2000: Environment is good for Business*. Paper presented to the Metrex Torino 2000 Conference.

THE CONTROLLOR AND AUDITOR-GENERAL 1999: *Report on Auckland City Council: Management of the Britomart Project*. Wellington: Controllor and Auditor-General.

THORNLEY, A. 1988: Planning in a cool climate. *The Planner*, July, 17–18.

THRIFT, N. 1986: The geography of international disorder. In Johnston, R. J. and Taylor, P. (eds), *A World in Crisis? Geographical Perspectives*. Oxford: Blackwell.

TOWLE REAL ESTATE 1987: *Fall 1987 Office Up-date: Internal Report*. Minneapolis: Towle Research Department.

TOWN AND COUNTRY PLANNING ACT 1977. *The Statutes of New Zealand, 1977*. Public Act 121. Wellington: Government Printer.

TOWN PLANNING ACT 1926. *The Statutes of New Zealand, 1926*, Public Act 52. Wellington: Government Printer.

TURNBULL, L. 1999: *Sydney: Biography of a City*. Sydney: Random House.

TUROK, I. 1992: Property-led regeneration: panacea or placebo? *Environment and Planning* A, 24, 1263–79.

WARD, S. V. 2002: *Planning the Twentieth-Century City: the Advanced Capitalist World*. Chichester: Wiley.

WARF, B. 1994: Vicious circle: financial markets and commercial real estate in the United States. In Corbridge, S., Thrift, N. and Martin, R. (eds), *Money, Power and Space*. Oxford: Blackwell.

WATES, N. 1976: *The Battle for Tolmers Square*. London: Routledge & Kegan Paul.

WATSON, S. and GIBSON, K. 1995: Postmodern politics and planning: a postscript. In Watson, S. and Gibson, K. (eds), *Postmodern Cities and Spaces*. Oxford: Blackwell, 254–65.

WEBBER, P. 1994: Planning and accountability. *Australian Planner* 31, 216–20.

WEBSTER, F. 2000: *Re-investing Place: Birmingham as an Information City?* Paper presented to the People, Cities and the New Information Economy conference, Helsinki, 14–15 December.

WHEATON, W. C. 1987: The cyclic behaviour of the national office market. *American Real Estate and Urban Economics Association Journal* 15, 281–99.

WIEFFERING, E. 1986: City officials say development policy a boon to homeowners. *City Business*, August 6th, 11.

WILKS-HEEG, S. 1996: Urban experiments limited revisited: urban policy comes full circle, *Urban Studies* 33, 1263–79.

WILLIAMS, B. and MacLARAN, A. 1996: Incentive areas for urban renewal. In Drudy, P. J. and MacLaran, A. (eds), *Dublin: Economic and Social Trends – Volume II*. Centre for Urban and Regional Studies. Dublin: Trinity College, 43–6.

WILLIAMS, B. and SHIELS, P. 2002: The expansion of Dublin and the policy implications of dispersal. *Journal of Irish Urban Studies* 1 (1), 1–19.

WOOD, P. 1976: *The West Midlands*. London: David and Charles.

WOODCOCK, S. C. 1991: *The Sadler's Well Royal Ballet: Now the Birmingham Royal Ballet.* London: Sinclair-Stevenson Ltd.
ZUKIN, S. 1995: *The Cultures of Cities.* Oxford: Blackwell.

Index

Note: Figures and Tables are indicated by *italic page numbers*, notes by suffix 'n' (e.g. '144n(5)' means note 5 on page 144).